NATIONAL INTERESTS AND THE MILITARY USE OF SPACE

This book was produced under the auspices of the
Center for Science and International Affairs, Harvard University.

NATIONAL INTERESTS AND THE MILITARY USE OF SPACE

Edited by
WILLIAM J. DURCH

Ballinger Publishing Company • Cambridge, Massachusetts
A Subsidiary of Harper & Row, Publishers, Inc.

International Standard Book Number: 0-88410-974-7

Library of Congress Catalog Card Number: 84-9241

Printed in the United States of America

Library of Congress Cataloging in Publication Data

Main entry under title:

National interests and the military use of space.

"Produced under the auspices of the Center for Science and International Affairs, Harvard University"—P.
Includes bibliographical references and index.
1. Astronautics, Military—Addresses, essays, lectures.
I. Durch, William J. II. John F. Kennedy School of Government. Center for Science and International Affairs.
UG1520.N38 1984 358'.8 84-9241
ISBN 0-88410-974-7

CONTENTS

LIST OF FIGURES

LIST OF TABLES

FOREWORD

Twenty-seven years into the Space Age, we are all witness to the rapidity with which this domain has come under human domination. In this brief time there have been 3,000 space launchings. These have inserted 14,000 identifiable objects into earth orbit. All told, 9,000 have returned or disintegrated and somewhat more than 5,000 remain. On the average, nearly one object is lost and one more launching occurs each day.

These numbers can be compared to the number of ships that sail the open seas. Numerically speaking, near space has become populated by manmade objects in twenty-seven years to roughly the same extent as the open seas have in twenty-seven centuries, or in the five centuries since long ocean voyages became possible.

Just as the seas have been used from the beginning for both military and civilian purposes, so it is for space. Thus far the military investment in space and space-related capabilities, though highly relevant to current military budgets, has been low compared to what the future might hold. Thus far no battles have been waged in space or for dominion over space, nor has transit through space been used to wage war on earth. Thus far the military use of space has contributed greatly to the knowledge that prevents adversaries from misjudging each other and to the deterrence that restrains the escalating possibilities of new violence.

One can already see that the second quarter-century of the Space Age is filled with new prospects for the military use of space that will dwarf all that has gone before. The choices that will be made, either by informed and humane judgment or by a blind grab for near-term advantage, will go far to shape the world of the 21st century.

The work reported in this volume is a modest effort to explore the nature of the alternatives that lie before us so that foresight can inform the choices to be made. I hope that the readers will bring to the reading of this work the kind of desire to understand, and concern to explore the issues, that the writers brought to the writing of it.

Paul Doty
March 1, 1984

PREFACE

This book grew out of an effort, begun in the spring of 1982, to bring together a group of individuals who were interested in and knowledgeable about the politics and technology of the military use of outer space. Trends in national defense policy pointing toward a new competition in space-directed weapons and, potentially, a fundamental realignment of American strategic policy suggested the need for serious and sustained attention to space and security issues. The resulting Space Working Group first met, at the Center for Science and International Affairs (CSIA), in September 1982.

CSIA is a permanent research center of the John F. Kennedy School of Government, Harvard University, dedicated to research and education in the security and arms control field. Over the years, its working groups have addressed such issues as nuclear proliferation, energy and security, political instability, ballistic missile defense, and the verification of nuclear freeze proposals. The unstinting support of CSIA and its director, Professor Paul Doty, has made this volume possible.

This book focuses on the military use of space by the United States and the Soviet Union. It is aimed at a readership having some familiarity with security issues but presumes neither expertise in space issues nor a technical background on the part of its readers. Primarily policy-oriented, it delves into technology as necessary for background information or to support policy arguments.

A number of individuals, in addition to the contributors to this volume, were associated with the Space Working Group and contributed to structuring the group's approach to this complex subject. I would like, in particular, to thank Ashton Carter, Bill Criss, Charles Glaser, Steven Miller, and Michael Nacht. For sage advice on the manuscript I am grateful to Brad Dismukes, Carol Franco, and Larry Smith.

I would like to express my appreciation to Daryl Battin and Novella Wooten for their able assistance in administrative matters; to Chris Allen, for lightning-fast library help; and to Christine Lundblad and Mary Ann Wells, for their amazing efficiency with a sometimes cranky word processor. I would be remiss if I did not also thank the contributors who, in this age of electronics and chronic overcommitment, obligingly turned out draft upon draft, hunched over the keyboards of their word processing terminals.

Lastly, and most importantly, I would like to thank my wife, Jane, for her encouragement, her support, and her keen eye for avoidance behavior.

W. J. D.
Arlington, Massachusetts
January 1984

INTRODUCTION

William J. Durch

A quarter century ago, humanity began an historic journey beyond the ocean of air that sustains life as we know it and into the much vaster ocean of space. Though the purpose of the first artificial satellites was peaceful, the missiles used to boost them into space were built for another application: the conduct (should it come to that) of nuclear war. The specter of such a war still hangs over us. That it has not yet occurred owes as much to the military use of space as to any other human endeavor of the past twenty-five years. Reconnaissance satellites, used by the United States and the Soviet Union for more than two decades, have made nuclear arms control agreements possible, providing the hard evidence that substitutes for trust in superpower relations. To speak, then, of the military use of space as either a new or a necessarily ominous phenomenon is to ignore two decades of history.

Neither is the development of space weapons a new phenomenon. Before the Limited Test Ban Treaty of 1963, nuclear weapons were tested in space. In the late 1950s the United States researched several concepts for the inspection and possible interception of space vehicles, as well as the X-20 Dyna-Soar, a manned orbital bomber/reconnaissance vehicle. Those programs were cancelled before they reached the testing stage. From 1964 through 1968, however, the United States did conduct thirteen unarmed antisatellite (ASAT) tests using

Thor missiles based at Johnston Island in the South Pacific. Their principal targets in wartime would have been Soviet orbital bombs, which would have passed near Johnston Island on the way to North America.[1]

H-bombs in orbit topped the list of space weapon concerns in the late 1950s and early 1960s, when the last great spate of writings on space politics and arms control occurred. The H-bomb itself was new, not five years old as a deliverable weapon. The intercontinental ballistic missile (ICBM) was newer still. In combination, they conjured up images of orbital 100-megaton weapons, ready for detonation without warning. Such weapons seemed to some only a matter of time. Present-day issues like space-based ballistic missile defense received little attention in early 1960s writings, and ASAT weapons were virtually unmentioned.[2]

In the late 1960s, the first Soviet tests of a nonnuclear ASAT, and testing of a "fractional orbit bombardment system," marked the second round in the "weaponization" of space. Testing continued through 1971 and then stopped. The United States ceased testing its Thor system in 1968 and dismantled it in 1975. Soviet ASAT tests resumed after a four-year hiatus in 1976. Within a year, the United States started development of its own nonnuclear ASAT. At no time, however, did either country pursue antisatellite or space-based weapons on anything resembling a crash basis.

Under the leadership of the Reagan administration that attitude may be changing, and a third round of space weapon development may be about to begin. The early 1980s have seen a growing interest in ASAT and space-based ballistic missile defense (BMD), both pro and con. The notion of using space-based weapons for BMD has caught the attention of some policymakers uncomfortable with the current strategic standoff and the arms control agreements that support it, the 1972 U.S.-USSR Anti-Ballistic Missile Treaty in particular. The ABM Treaty prohibits, among other things, the development, testing, or deployment of space-based ABM systems by either superpower. President Reagan's 23 March 1983 "star wars" speech, and follow-up statements by Secretary of Defense Weinberger, in essence question the wisdom of that prohibition.[3]

ASAT systems using 1980s technologies will mature in the next five years. Such technologies—compact and proliferated—may pose substantial problems for arms control once tested and operational, problems of verification in particular. To those who see a virtue in controlling ASAT weapons, this lends a sense of urgency to the issue.

There is, in short, good reason to examine carefully at this juncture both the premises and promises of the military use of space.

Chapter 1 introduces readers to the environment of space, provides basic information on rocketry and orbits, and describes current U.S. and Soviet space launch sites and launch vehicles. It concludes with a discussion of the potential future role of humans in space.

Chapters 2 and 3 discuss the American and Soviet space programs, respectively. Chapter 2 emphasizes the evolution of American military policy and capabilities. Chapter 3 analyzes the contemporary Soviet space program and addresses, in particular, the high Soviet launch rate and its implications for American security interests.

Satellites are fragile objects. Chapter 4 examines the ways in which they can be prevented from carrying out missions, from direct attack through disrupted communications, as well as ways to protect them.

Protecting the fleet from satellite-targetted attack is one reason given for an American ASAT program. Chapter 5 examines the relationship between space and naval operations.

Chapter 6 looks at "star wars." It discusses, in particular, the technical feasibility, military utility, and strategic implications of space-based laser and particle beam weapons.

Further arms control measures for outer space would not be negotiated in a legal vacuum. Chapter 7 traces the evolution of the international law of outer space and examines both current agreements and current disputes. The evolution of Soviet attitudes toward space issues, as reflected in the writings of its most prominent jurists, is the subject of Chapter 8.

Future arms limitation measures for outer space could range, theoretically, from "rules of the road" for space objects to complete demilitarization. Chapter 9 examines the range of possible approaches, giving pros and cons for each.

This volume concentrates on the bilateral superpower relationship. There are other aspects of the military use of outer space that considerations of print space, and time, preclude treating extensively. They include the relationship between military and civilian uses of space, the ownership and use of satellites by increasing numbers of states and nongovernmental entities, and the impact of domestic politics and institutional interests on the shape and direction of the American space program. These and other factors contribute to the complexity of issues that are only likely to grow more complex as the years go by.

Sorting out policies for complex issues requires some kind of sorting mechanism. Two mechanisms that make the task easier are mental images and reasoning by analogy—assumptions about what something is like and links made to previous experience. Looking at images and analogies for space will set the stage for the chapters that follow.

VISIONS OF SPACE

There are emotional underpinnings to the recent upsurge of interest in the military use of space. Romantic notions of space—imbued in one generation by Flash Gordon serials, nurtured in the next by the Apollo moon program, and sustained, larger than life, by *Star Trek* and *Star Wars*—are like those romantic notions of China reinforced by every VIP puffing up the Great Wall, as impervious to data as Deng Hsiao-Ping is to Mao-thought. To some, the "high frontier" beckons, and its call is a siren song: Space and space weapons will save us from nuclear war. To others, space is a separate ecology to be saved from military depredation. This image of apartness is shared but inverted by some space-weapons advocates, who see in space an arena in which to engage the enemy, show resolve, and signal intent, with little risk of terrestrial spillover. Space is a sanctuary from the military, on the one hand; for the military, on the other.

The reality content of both these images is reasonably low. All current military space systems support terrestrial military operations. That will continue to be the case, so space will not be immune from the effects of combat originating on earth. Direct attack ASATs may or may not be available for use in a future conflict—and if available may or may not be used—but electronic countermeasures will surely be brought into play since most exchanges between earth and space assets are electromagnetic in nature. Moreover, conflict originating in space is unlikely to remain confined there, especially if one combatant power sees space as an arena for separate, symbolic conflict, and the other does not; or if one considers attacks on space objects to be not very significant, while the other sees them as harbingers of escalation.

Some view outer space as the ultimate high ground. On land, the holder of high ground has a better view of the surrounding countryside, and gravity assists the hurling of his projectiles and retards those of the enemy. On the other hand, to the extent that it is useful for surveillance, high ground is also exposed. To the extent that the other side's weapons can shoot higher than the top of the hill, having gravity on one's side may not matter all that much. Moreover, gravity itself becomes an enemy when the time comes to reinforce or resupply.

Space power advocates periodically assert that the country that claims the high ground of space, that climbs farthest and fastest out of earth's gravity well with full military capabilities, will secure for itself a commanding military position. That vision remains unfulfilled. The United States and the Soviet Union have been content thus far to share the high ground, in part because both benefit (as in the mutual use of photoreconnaissance satellites to monitor the other side's forces and compliance with arms control agreements), and in part because the economic, military, and political costs of enforcing a regime of space dominance would be prohibitive.

Though the military use of space is not readily separable from terrestrial military policy and operations, space is a different environment, and different rules govern the behavior of objects moving through it. It presents some unique advantages and poses some unique hazards to people and their equipment. Although it may be "high," it is not "ground." Indeed, the threat to clear thinking about policy posed by overly romantic and unreal images may hold more potential for harm to national security than any actual space threat the country now faces.

THE OTHER FRONTIERS

In tracing and analyzing the military use of space, it may be helpful to have in mind earlier human experiences with other "new frontiers." Space is the latest environment to come within reach of human activity—as impelled by advancing technology, political rivalry, economics, or military necessity—but it is hardly the first.

Humanity may have taken as much as one thousand years to progress from river rafts to intracoastal trade, and stayed basically within sight of land for many thousand more. The technologies of open-ocean sailing (the compass, sextant and full-rigged ship) came together only in the late fifteenth century, and only in the mid-1800s were sailors freed from the whims of the wind. The conquest of the air began in 1783, when the Montgolfier brothers flew their first balloon over Paris; it was complete within two centuries. The polar regions resisted human exploration for the better part of three centuries.

Use of space has developed so quickly, by contrast, that we often think about space by analogy to these earlier frontiers. Analogies to the military use of sea and air abound, and space law owes much to usages evolved many years before the launch of Sputnik I.

Military Use

Naval forces developed to protect seaborne trade from piracy and coastal states from seaborne attack. Fleets were also important adjuncts of invading land armies, supplying food and other necessities more cheaply and conveniently than could be done by land. This dependency made some armies vulnerable to loss of sea-based support. Greek destruction of the Persian fleet at Salamis (400 B.C.), for example, forced the retreat of a Persian army too large to live off the Greek countryside.[4]

Naval battles have been the pivots of history more recently as well. Nelson's victory at Trafalgar (1805) assured British naval supremacy and forestalled invasion of Britain by Napoleon. American victory at Midway (1942) irrevocably reversed Japanese fortunes in World War II.[5]

Parallels are readily drawn to military use of space. Satellites provide certain kinds of support to terrestrial forces more cheaply and conveniently than other available means. Indeed, some suggest that they are, like the Persian fleet at Salamis, so vital to wartime operations that U.S. commanders would be as stranded as Darius was should access to space be lost. Others suggest that fleets of orbiting battle stations could, like their seaborne predecessors, play an historically decisive role in a future East-West war.

Analogies to airpower are common as well. Aircraft first played a major military role in World War I. Used initially against observation balloons, fighters soon spent most of their time engaged with one another. By the close of the war, they had been joined by huge, lumbering biplane bombers that inspired an Italian officer, Giulio Douhet, to pen the first theory of strategic airpower in 1921.[6]

Douhet imagined the next major war as one in which great fleets of bombers would disable the enemy with poison gas and high explosives before ground armies ever moved into action. His theories were not borne out in World War II. Poison gas, admittedly, was not used. But relentless allied bombing raids into Germany did little to sap that country's will to resist, and Gen. Clair Chennault's 14th Air Force did little to stop the Japanese ground advance through China (Chennault's bases kept being overrun by the forces they were supposed to be terrorizing). Some believe that Douhet merely underestimated the necessary destructive power per aircraft. In the nuclear age that is no longer a problem.[7]

There is a kinship between Douhet and modern advocates of space power. Just as Douhet's bombers were to start and finish a war before traditional elements of power could be brought into play, so space battle fleets are advocated to nip a nuclear war in the bud, hosing enemy missiles with irresistible fire.

Antarctica was the last terrestrial frontier. Never a target of military planning, the remotest and most hostile region of the world, it was sealed against military activities in 1959. Though it is sometimes offered as a model for space, Antarctica has never offered military advantages that exceed the costs it imposes. Any military activity based there could be carried out more efficiently from some other point on the globe. The arguments for many military uses of space, however, are cast in just such cost-effectiveness terms, making Antarctica, in that sense, not the analog but the inverse of space.

International Law

The relative suddenness of the push into space has meant that its orderly development has been largely a function of positive rather than customary law; that is, of treaties and agreements rather than rights secured by long practice. Its major concepts are derivative of the laws of sea and air, and their competing notions of freedom and sovereignty.

Freedom of the seas—their openness to use by all and appropriation by none—was a notion developed in the seventeenth century by Dutch jurists to justify equal ocean access for Dutch merchants. It proved so successful an instrument of Western expansion (enforced into the twentieth century by the British Royal Navy) that it has been seriously challenged only in the last few decades, as sovereign states that are not major ocean users have proliferated. Increasing numbers of these states, joined by many of the major powers, have claimed 200-mile coastal resource zones. Ocean transit through such zones is nominally unaffected, but there remain concerns about "creeping jurisdiction"—the transformation of economic zones into territorial seas, zones of complete national sovereignty.

Applied to space, freedom of the seas has meant freedom of access and use. Anyone can put a payload into orbit, provided it does not interfere with other space objects. No international organization grants permission for launch, and no state can veto another's activities.

Just as freedom of the seas extends to the air above the oceans, sovereignty extends to the air above national territory. Legal scholars began to think about safeguarding national interests against airborne incursions around the turn of the century. Early concerns involved espionage, not bombardment (balloons had long been used for wartime observation). By 1913, the International Law Association asserted states' rights to restrict the passage of aircraft over their territories and territorial waters. World War I and the growth of military aviation stimulated universal acceptance of the concept. In 1919, the Paris Convention (for the Regulation of Aerial Navigation) gave states complete and exclusive sovereignty over their airspace. A separate agreement, signed at Chicago in 1944, reaffirmed the concept.[8]

The breadth of national airspace is defined by the outer limit of the territorial sea and tends to be vigorously defended against intrusion. The upper reaches of national airspace are defended as well, but their limit is undefined. In principle, airspace is infinite; in practice, it appears that craft that enter orbit exit national airspace. Satellites routinely overfly states' territories collecting data, sometimes at relatively low altitude, without major military incident.

Analogies are useful, but they are never exact. Space is neither sea nor air nor ice. Yet the record of human experience in dealing with these other environments suggests what we can expect for space. It is not yet as accessible as the other three—the technology threshold is higher. But more countries are crossing the basic threshold, and the superpowers stand at the next one. Whether they cross it will depend in part on their assessments of the cost and relative military utility of doing so. In the past, that which has had military utility has generally been undertaken; limitations, negotiated or self-imposed, have been the exception.

The chapters that follow present information and arguments about past, current, and prospective military uses of space, space law, and approaches to space arms control. Each chapter reflects the assessments and the conclusions of its author(s). There is no deliberately orchestrated point of view for the collection as a whole. It should also be stressed that the views expressed by the authors do not necessarily reflect the views of the organizations with which they are affiliated nor the views of their fellow contributors.

NOTES

1. Bhupendra Jasani, "Space: Battlefield of the Future?" in Michiel Schwartz and Paul Stares, eds., *Space: Past, Present and Future,* a special issue of *Futures* (October 1982), p. 437; and U.S. Congress, House, Committee on Science and Technology, *U.S. Civilian Space Program, 1958–1978, Vol. I.* Committee Print, Serial D, January 1981, App. G.
2. See, for example, Lincoln Bloomfield, ed., *Outer Space: Prospects for Man and Society* (Englewood Cliffs, N.J.: Prentice-Hall, 1962); and Joseph M. Goldsden, ed., *Outer Space in World Politics* (New York: Praeger Publishers, 1963).
3. For the text of the President's speech, see *New York Times,* 24 March 1983; Secretary Weinberger spoke on *Meet the Press,* Betty Cole Dukert, producer, 27 March 1983. (Mimeo.)
4. William Stevens and Allen Westcott, *A History of Seapower* (Garden City, N.J.: Doubleday and Co., 1942), pp. 10, 17.
5. Ibid., p. 122; Michael Lewis, *The History of the British Navy* (Baltimore: Penguin Books, 1962), p. 194–95; and Kenneth Davis, *Experience of the United States in World War Two* (Garden City, N.J.: Doubleday and Co., 1965), p. 233.
6. Bernard Brodie, *Strategy in the Missile Age* (Princeton: Princeton University Press, 1965), p. 72.
7. Ibid., p. 82–87; Barbara Tuchman, *Stillwell and the American Experience in China* (New York: Bantam Books, 1971), pp. 429–32, 584–86.
8. D.H.N. Johnson, *Rights in Air Space* (Dobbs Ferry, N.Y.: Oceana Publications, 1965), pp. 12, 21, 33, 58.

REFERENCES

Bloomfield, Lincoln P., ed. *Outer Space: Prospects for Man and Society.* Englewood Cliffs, N.J.: Prentice-Hall, 1962.

Brodie, Bernard. *Strategy in the Missile Age.* Princeton: Princeton University Press, 1965.

Davis, Kenneth. *Experience of War: The United States in World War Two.* Garden City, N.J.: Doubleday and Co., 1965.

Dukert, Betty Cole, producer. "Caspar W. Weinberger, Secretary of Defense." *Meet the Press,* 27 March 1983. (Mimeo.)

Goldsden, Joseph, M., ed. *Outer Space in World Politics.* New York: Praeger Publishers, 1963.

Jasani, Bhupendra. "Space: Battlefield of the Future?" In M. Schwartz and P. Stares, eds. *Space: Past, Present and Future.* A special issue of *Futures* (October 1982).

Johnson, D.H.N. *Rights in Air Space.* Dobbs Ferry, N.Y.: Oceana Publications, 1965.

Lewis, Michael. *The History of the British Navy.* Baltimore: Penguin Books, 1962.

"President Reagan's Speech on Military Spending and a New Defense." *New York Times,* 2 March 1983.

Stevens, William and Westcott, Allen. *A History of Seapower.* Garden City, N.J.: Doubleday and Co., 1942.

Tuchman, Barbara. *Stillwell and the American Experience in China.* New York: Bantam Books, 1971.

U.S. Congress. House. Committee on Science and Technology. *U.S. Civilian Space Programs, 1958–1978, Volume I.* Committee Print, Serial D, January 1981.

1 STEPS INTO SPACE

William J. Durch & Dean A. Wilkening

Space has awed humanity ever since people first noticed the stars. Lacking explanations for what they saw in the heavens, ancient peoples attributed the passage of comets, the rising of the sun, and the setting of the moon to the actions of mysterious, powerful beings who ruled the world and all things in it by a combination of art and caprice. Sky gods—hurlers of fire and brewers of storms—are common to virtually all ancient mythologies.

But the sky was more than a source of myth; it was a source of measurement. From the regular movements of objects in the heavens, priests of the ancient river—valley civilizations could predict the coming of spring and the annual flood, and direct the planting of crops (and the sacrifice of kings and maidens). The astronomers of ancient Babylon had, by the eighth century B.C., compiled detailed charts of the movements of the stars and planets and were able to predict the precise dates of, among other things, solar and lunar eclipses. Knowing when the sun-god would withhold his light, and "causing" him to give it back, would have been a source of tremendous political and religious power for the priesthoods.[1]

The discovery that the entire observable universe behaved according to predetermined patterns led to the inference that human behavior was itself so determined and, as part of the great mechanism of the universe, could be predicted by the same starry indicators.

Astrology still wields great influence over the lives of some people and cultures.[2]

Greek astronomers were the first to suppose that the moving lights in the sky might be other physical objects. But the Greeks, like their predecessors, believed the universe to be geocentric (earth-centered) and spent much of their energy trying to make that model fit the motions they observed. One astronomer, Aristarchus of Samos, argued for a heliocentric (sun-centered) universe as early as 280 B.C., but his views were much in the minority.[3]

The best-known publicist of the geocentric view was Claudius Ptolemy, a resident of Alexandria, Egypt, in the second century, who summarized then-available knowledge of astronomy. Transmitted to Europe via the Arab world, his *Almagest* and the geocentric universe it described fitted well with medieval Christian concepts of man as the center of God's creation.

It was the last major work on the subject for 1,000 years. Until the late thirteenth century, the church forbade the notion that there could even *be* more than one world, and Ptolemaic theory acquired the force of dogma. When renewed scientific inquiry raised doubts about that universe, the response of the church (both Catholic and Protestant) was to condemn the doubters. Nicholas Copernicus's *On the Revolution of the Celestial Orbs* (1543), the first work in Western civilization to postulate a sun-centered universe, was placed on the Catholic church's Index of Forbidden Books in 1616, where it remained until 1835. Galileo, whose work confirmed that of Copernicus, was forbidden to teach it; his case was not formally reopened by the church until 1982.[4]

As the instruments of astronomy improved, it became clear that even the sun was not the center of existence, but a minor object in the cosmos, the hub of its own system but an outlier on the spiral arm of an average-sized galaxy, one of billions of similar galaxies speeding away from one another into the depths of a universe with no discernible end.

THE SPACE ENVIRONMENT

Neither has space a discernible beginning. Ninety-seven percent of the earth's atmosphere lies below 30 kilometers altitude, the approximate limit of the stratosphere. By the time one reaches 80 kilometers, atmospheric pressure is down to a millionth of what it is at sea level, and at 160 kilometers to less than a billionth.[5]

Near-earth space is a far from benign environment. Data from the first U.S. satellites showed the earth to be all but surrounded by high energy particles—protons, electrons, and hydrogen and helium nuclei—trapped in the earth's magnetosphere. The magnetosphere starts about 600-1,000 kilometers above the earth's surface and collects particles of both cosmic and solar origin. The "solar wind" constantly showers the earth with particles moving at an average of 500 kilometers a second; the occasional solar flare unleashes additional electrons and photons moving as fast as 2,000 kilometers a second.[6]

The radiation belts, named after their discoverer, Dr. James Van Allen, reach out into space roughly 64,000 kilometers. The magnetosphere curves in toward the surface at the poles, so the Van Allen belts extend only to about 75° north/south latitude on the sunlit side of the planet and somewhat less on the dark side. Electronic instruments and people passing through the belts have to be shielded against their radiation. (Early satellites in geosynchronous orbit, in the heart of the outer belt, developed electrostatic charges as high as 15,000 volts.) The radiation problem received much attention in the Apollo moon program, and even today it is a rare long-duration spaceflight that carries its crew higher than 500 kilometers. Solar flares pose an even greater radiation hazard, and their impact varies little with orbital altitude.[7]

Heat and cold are also major problems. Heat from the sun—unshielded by the atmosphere—and from a spacecraft's own electronic components must be dissipated. This was a major problem for the U.S. Skylab, which lost a part of its heat shielding on launch in 1973. The first crew rigged a gold-foil parasol to cover the damaged area, and the Skylab mission went forward. Spacecraft in shadow (of the earth or moon, for example) must contend with the opposite problem. Though the widely dispersed particles in space may have high kinetic temperatures of their own, a thermometer waved through space and shielded from the sun would register −270 degrees Celsius, or very near absolute zero. Thus, spacecraft must be designed to deal with two thermal extremes, a difficult problem if human beings are involved since humans work efficiently only within a very narrow temperature band.[8]

THE FIRST MEN IN THE MOON

Humans were involved first in spaceflights of fancy. In A.D. 160, the Greek satirist Lucian penned a *True History* of Odysseus' "missing adventure" among the inhabitants of the moon. Fourteen hundred years later, Dutch astronomer Johannes Kepler's *Somnium* related the adventures of earthly travelers borne to the moon by spirits. His space transportation system notwithstanding, Kepler's account of conditions in space and on the moon (except for its population) was remarkably accurate. He deduced the lack of air in space or on the moon from his calculations of planetary motion, which did not permit the friction that such an atmosphere would produce.[9]

In the nineteenth century, two other writers of space fiction, Jules Verne and H.G. Wells, contributed to the conquest of space by firing the imaginations of young scientists. Verne's *From the Earth to the Moon,* published in 1865, used an immense cannon to fire a manned capsule around that body. Wells' *First Men in the Moon,* published in 1901, makes use of an antigravity paint to send its voyagers on their way. These and other fictional accounts showed no lack of imagination, but all reverted to fanciful means of propelling their characters for lack of a technically feasible means of doing so. Verne's cannon sounds plausible until it is realized that the acceleration needed to boost the capsule to escape velocity (about 40,000 kilometers an hour) before leaving the gun tube would crush the human body. The basis for more practical means of spaceflight was provided by two scientists inspired, as young men, by the work of Verne and Wells.

Konstantin Tsiolkovsky (1857–1935) worked out the mathematics of rocketry (the "rocket equations") and published them the same year the Wright Brothers flew at Kitty Hawk (1903). He developed the principle of "staging" (the discarding of spent fuel tanks and motors) and deduced the best fuel combinations for high performance liquid rocket engines, including liquid oxygen/liquid hydrogen. His report was reissued by the Soviet government in 1923, as a wave of interest in rocketry swept Europe.[10]

Robert Goddard (1882–1945) was a theoretician and inventor who developed the first successful liquid-fueled rocket, powered by liquid oxygen and gasoline. A loner, he refused to cooperate with other, larger organizations – such as Cal Tech – whose resources might have furthered his work. Fearful that others would copy his work to his disadvantage, Goddard published little—a brief report to his funder, the Smithsonian Institution, in 1919, and an equally brief paper on liquid rocket propulsion, in 1936. To protect his work, Goddard filed many patent applications and ultimately held more than 214 rocket-related patents. During World War II he helped the U.S. Navy develop rocket-assisted takeoff units for military aircraft.[11]

ROCKET BASICS

The fundamental principle behind rocketry is the conservation of linear momentum. When hot gases are expelled from the exhaust nozzle of a rocket, the rest of the rocket's mass reacts in an equal and opposite fashion, moving the rocket forward. (It does not "push" against the air; in fact, a rocket engine works more efficiently in the vacuum of space than it does within the atmosphere, where the rocket body is slowed by atmospheric friction.) In a chemical rocket, the type used in all space launch vehicles, the burning of fuel and oxidizer in the combustion chamber produces high temperatures and high pressures. The higher the temperature and pressure, and the lighter the molecules in the exhaust gases, the higher the exhaust velocity.[12]

Thrust is a commonly used term that measures the amount of force a rocket engine can generate, and thus how fast the rocket will accelerate. It is proportional to the exhaust velocity and to the rate at which fuel is consumed. To launch a rocket against earth's gravity, its thrust must at least equal its total weight. A rocket with a thrust-to-weight ratio of less than one will sit on its launch pad happily consuming fuel—the bulk of its weight—until the thrust is finally greater than the remaining weight, at which point it will take off on an abbreviated flight. (If it does not have to fight gravity, a rocket motor can do useful work even if its thrust is much less than its vehicle's total mass. Small attitude thrusters on orbiting spacecraft are good examples of such motors.)

A rocket accelerates as long as its engine burns. Assuming it burns continuously, maximum speed is reached when its fuel is exhausted ("burnout"). The change in velocity between rocket ignition and burnout is called ΔV ("delta vee"). For a single stage rocket the equation for ΔV, ignoring the effects of gravity, is

$$\Delta V = v_e \ln[M_o/M_b],$$

where v_e is the exhaust velocity, ln[ ] is the natural logarithm of the term contained in the brackets, M_o is the initial mass of the rocket, and M_b is its mass at burnout. This is the basic "rocket equation." To achieve large velocity changes, rocket designers strive for high exhaust velocities and a large "mass ratio" (M_o/M_b), which results when a very large fraction of the initial mass is fuel.

The propellant in a chemical rocket may be either liquid or solid. Each type has advantages for particular applications. There are two basic types of liquid propellants: cryogenic and hypergolic. In the first type, one or both components are normally gases. They must be condensed into liquids (at very low temperatures, minus several hundred degrees Celsius) so that the fuel can be stored compactly. Liquid oxygen (LOX) was the first cryogenic material used—in 1926 by Robert Goddard. Over the years it has been combined with various fuels (gasoline, alcohol, kerosene) and ultimately, in the early 1960s, with liquid hydrogen (LH_2—the combination Tsiolkovsky had recommended, correctly, as the highest-energy rocket fuel). LH_2 is even colder and more difficult to handle than LOX, and because it is such a light element it will seep through most containers. Even so, its advantages are such that hydrogen/oxygen engines now power a number of boosters, including the space shuttle orbiter.

The term "hypergolic" is derived from the Greek meaning "excess energy." Hypergolic fuels and oxidizers burn on contact, without an external source of ignition. Since they are not cryogenic, they can be stored in a booster's tanks for a long time, ready for use on short notice. Thus, hypergolic propellants have long been used in military missile programs and restartable upper stages.

Solid fuel rockets are simpler in principle than their liquid fuel counterparts, requiring no fuel pumps or cryogenic handling. This tends to give them high reliability. In practice, however, the chemistry of solid propellants is quite complex. Since they are commonly used in military applications, technical advances in solid fuels are often closely held secrets.

Solid propellant is formed into a cylindrical shape called a grain, in which the fuel and oxidizer are mixed. When ignited, the propellant burns inside a cavity running along the axis of the grain. The propellant is, in effect, its own combustion chamber.

Solid rockets burn their propellant at a high rate and can be fabricated in large sizes (an engine test-fired in the 1960s had a diameter of 6.6 meters). Thus, they can be made to generate thrusts equal to or greater than those of liquid rockets.[13] Current U.S. launch vehicles (discussed below) capitalize on the advantages of both types of rocket motors, using solid boosters to augment takeoff thrust and liquid-fueled engines to reach orbit, to maneuver, and to reenter.

WAR ROCKETS

The military rocket is a device whose pedigree is obscure. Though many credit the Chinese with their first use in the thirteenth century, there is some indication that the formulae for the propellants used in those rockets may have come to China from Europe. On the other hand, the Mongol expansion of the middle thirteenth century may have transported Chinese technology westward. That same expansion brought rocketry to India, where it was encountered by the British as early as 1750. Indian war rockets were used primarily to spook cavalry (in effect, as early jamming devices), and at that they were apparently effective.[14]

William Congreve (1772–1828) drew upon the British experience in India in designing his war rockets (the "rockets red glare" of the "Star Spangled Banner"). Congreve's rockets came in various sizes and flew with the aid of a trailing stick/stabilizer. An eight-inch Congreve rocket weighted about 135 kilograms, carried a 23-kilogram explosive warhead, and had an effective range of about two and a half kilometers. Some had twice the range and accuracy of early nineteenth century artillery and came into widespread use with the Royal Navy. But the guidance stick put clear limits on their range and accuracy, and, as artillery improved, Congreve rockets became obsolete. Readoption of the rocket for military purposes awaited improved propellants and guidance. Both were available by the 1930s. Had the U.S. government been more interested and Goddard less fearful of rivals, the United States could have been a world leader in the field. In the event, that title fell to Hitler's Germany.[15]

The Treaty of Versailles that ended World War I effectively disarmed Germany but, like many other arms control and disarmament agreements, it did not cover all potential weapon systems. In 1919 few thought of the rocket as a potentially practical weapon of war, and the treaty did not mention them. In 1923, Hermann Oberth published *The Rocket Into Interplanetary Space,* a seminal work that led to the founding of rocket societies in Germany, the United States, Britain, and Russia (and occasioned the republication of Tsiolkovsky's work). The German organization, the Society for Space Travel [*Verein fur Raumschiffart* or VfR], was founded by Oberth himself in 1927 and launched its first small liquid-fueled rocket in 1931.[16]

The VfR's work came to the attention of the German Army, which had instituted a modest rocketry program of its own in 1929. Still operating under the restrictions of the Treaty of Versailles (Hitler had not yet come to power), the army was interested in the rocket's potential as a substitute for conventional artillery. The young Wernher von Braun, working with the VfR, seemed particularly able. He soon became the first civilian employee of the fledgling army rocket program; his PhD dissertation on combustion processes in liquid propellant rockets, completed in 1934, was done for and classified "secret" by the government.[17]

In April 1937, military rocket research moved to the Baltic test station of Peenemunde, in a venture jointly (and jealously) funded by the Wehrmacht and the Luftwaffe. Von Braun's research led ultimately to 1,000 ballistic missile attacks on London in the final six months of World War II.[18] That missile, the V-2, was father of the battlefield rockets (American and Soviet) of the early 1950s, and grandfather of the nuclear-tipped intercontinental ballistic missile, or ICBM. That class of missile became, in turn, the workhorse of the space age.

REACHING FOR SPACE

The Soviet Union was the first country to place a man-made object in orbit around the earth. That feat should not have come as a surprise and need not have been a first. On 10 January 1955 Radio Moscow announced that the launch of an artificial earth satellite was possible "in the near future," and that August the Soviets stated their intent to orbit a satellite during the planned International Geophysical Year (1957–58). In October 1956 *Aviation Week* magazine published excerpts from an article in *Moscow News* that stated that the anticipated Russian satellite would be 24 inches in diameter, weigh over 100 pounds, and fly a 185 by 810-mile orbit (the actual parameters of Sputnik I were 23.8 inches, 184 pounds, and 156 by 560 miles). In June 1957 the Kremlin reported to the IGY headquarters that their satellite project was ready; in September they broadcast the transmission frequencies it would use; and on 4 October they launched it, to the consternation of the American public and many of its leaders.[19]

It is understandable that such pronouncements from Moscow should have been discounted. This was the same era in which Soviet leaders flew their handful of bombers past reviewing stands time after time, creating a "bomber gap" panic in the West. It was a time in which boasting and disinformation from Moscow were expected, and straight news was not. Not attending to Soviet cues caused needless shock and surprise, but that is not why America was second into space.

In May 1946 the Rand Corporation estimated that a 400-pound vehicle could be put into orbit by 1951, given a concerted effort to do so.[20] The Air Force, for which Rand worked, was only modestly interested in space. The size and weight of then-available nuclear weapons would have required huge, unwieldy missiles to carry them intercontinental distances, making missiles a poor second choice to jet bombers, which the Air Force busied itself building in quantity. No missiles meant no spacecraft.

The Army, by contrast, had great interest in the development of strategic missiles as a means of getting a piece of the nuclear action. Von Braun and his team at the Army Ballistic Missile Agency, Huntsville, Alabama, set about developing those missiles. Their Redstone, first tested in 1953, was basically a scaled-up V-2. In the spring of 1955, Von Braun proposed that a Redstone-derived vehicle be used to launch an American satellite into orbit. Project Orbiter might have produced a satellite as early as the summer of 1956 had it been given the go-ahead. But the United States chose not to use a military rocket as its launch vehicle, inasmuch as the project was to be conducted under the peaceful auspices of the IGY. Project Vanguard received the nod, and Orbiter was officially terminated.[21]

In an organization as complex and sprawling as the Department of Defense, an occasional order fails to be executed fully. Orbiter continued to exist as an adjunct to the Jupiter intermediate range ballistic missile program. Dubbed Jupiter-C, in September 1956 a four-stage Orbiter vehicle flew 5,300 kilometers and almost reached orbital velocity (had its fourth stage not been a dummy, it could have ended up in orbit). Washington immediately warned the Army against launching any "accidental satellites," which would, of course, preempt Vanguard.[22]

That project ran afoul of numerous technical problems. The

launch of Sputnik I prompted rescheduling an early test vehicle as the real thing, and in December 1957, with live television coverage and the prestige of the United States on the line, that hapless vehicle blew itself up.[23]

The Army was given permission to prepare a Jupiter-C for launch. Explorer I was sent into orbit on 31 January 1958, eighty days after Orbiter was officially reborn.

ORBIT BASICS

When an object is accelerated to a speed of eight kilometers per second (km/sec) along a path tangent to the earth's surface, at an altitude of about 100 kilometers, its tendency to fly off in a straight line is just balanced by the force of gravity pulling it back. In the low-friction environment of space it continues to fall around the planet; that is, it enters into orbit.[24]

Atmospheric drag roughly below 100 kilometers is so severe that an object quickly loses velocity and falls back into the atmosphere; unless designed for reentry (or very massive), friction heating incinerates it completely. Large satellites and those with solar panels (which act like air brakes) will decay faster and from higher altitudes. Skylab, for example, decayed from an orbit of more than 400 kilometers. Satellites with on-board maneuvering capability can boost or otherwise alter their orbit from time to time, to the limit of their fuel supplies. Such maneuverability is especially important for reconnaissance satellites, enabling them to be used in a more flexible and timely fashion.

Orbits vary from near-circular to highly elliptical (that is, vary in *eccentricity*). At one end of an elliptical orbit a satellite reaches its highest altitude (*apogee*) and its lowest velocity (it has been climbing away from the earth, against the force of gravity). As the satellite loops back toward the planet, gravity acts with increasing force, so maximum orbital velocity is reached at the opposite end of the ellipse, which is also the lowest altitude (*perigee*). The more nearly circular the orbit, of course, the more constant the orbital velocity.

The *period* is the time it takes a satellite to complete one circuit of its orbit. Period varies according to the orbit's width (the straight-line distance from apogee, through the earth, to perigee). Thus a satellite in a circular orbit at 20,000 kilometers altitude will have about the same period (twelve hours) as a satellite in an eccentric orbit with a 500 kilometer perigee and 40,000 kilometer apogee.

A space object's orbital plane (the disk outlined by the object as it traces its orbit) always cuts through the center of the earth. It may parallel the equator, pass through the poles, or lie somewhere in between. The angle between the equator and the orbital plane is the orbit's *inclination*. An equatorial orbit has zero inclination; a polar orbit ninety degrees. By convention, inclination is measured on a satellite's northbound pass over the equator (its *ascending node*). Most satellites are inclined at less than 90 degrees, and so move west to east (in the direction of the turning earth); objects in *retrograde* orbits, inclined at more than 90 degrees, move east to west.

A satellite's inclination defines the highest latitude crossed by its ground track—the projection of the orbital plane onto the earth's surface. A satellite in a low altitude orbit inclined 45 degrees, for example, will never pass over Leningrad (latitude 60° north) or the Falkland Islands (51° south). A satellite in polar orbit, on the other hand, will pass over every point on earth.

Combining these orbital parameters in particular ways produces orbits with different utilities. The plane of a satellite in low earth orbit (LEO) inclined 98 degrees, for example, changes by about one degree a day, such that the plane keeps the same orientation toward the sun throughout the year. The satellite continually passes over points on earth at the same local time. Such a *sun-synchronous* orbit is particularly useful for imaging, since objects seen on successive days or weeks will be seen under similar lighting conditions (important if one is tracking the progress of some activity or looking for telltale changes in an object).

In *semi-synchronous* circular orbit (altitude 20,200 kilometers) a satellite circles the earth once in twelve hours. The clockwork regularity of this pattern is well suited to position finding, a fact exploited by both the United States and the Soviet Union in their newest generations of navigation satellites.

Irregularities in the shape and density of the earth cause the *major axis* (apogee-perigee line) of eccentric orbits to rotate, like a slow-motion, planetary hula-hoop; not desirable if a satellite's task is communication with or surveillance of a particular part of the globe. At an inclination of about 63 degrees, however, the gravitational perturbations balance out and the orbit is stable. Soviet Molniya communications satellites, which reach apogee high above the Soviet Union, use a 63-degree inclination; the highly eccentric, twelve-hour orbit they trace is named for them. As Chapter 3 relates, the Soviet Union uses Molniya orbits a great deal.

Finally, there is the familiar *geosynchronous* orbit (GSO); circular, usually with near-zero inclination, at an altitude of 35,900 kilometers. With a period of twenty-four hours, a satellite at GSO moves with the earth and appears stationary from the ground. Geosynchronous orbit is thus ideally suited for communications satellites, both military and civilian.

To date, the chemical rocket is the only known means of reaching any of these orbits from the ground. The boosters used by the United States and the Soviet Union and the complexes used to launch them, are the subject of the next section.

LAUNCH SITES AND BOOSTERS

In the quarter century since Sputnik and Explorer I, more than 3,000 payloads have been sent into space, most into earth orbit, some to the moon, others to the planets. Pioneer 10, launched in March 1972, passed the orbit of Pluto in June 1983, on its way out of the Solar System.

Most of those payloads have been launched by the United States or the Soviet Union. Both employ military missiles as space launch vehicles. Each has a pair of principal launch sites and subsidiary sites used for minor missions and sounding rockets.

U.S. Launch Sites

Principal American sites are the Kennedy Space Center (KSC) at Cape Canaveral, Florida (latitude 28.5° north; longitude 80° west) and Vandenberg Air Force Base in southern California (latitude 32° north; longitude 120° west). Payloads destined for equatorial or other low-inclination orbits are launched from Kennedy out over the Atlantic. West-to-east launches are the most energy-efficient because they take advantage of the earth's spin—about .47 km/sec at the equator and .41 km/sec at KSC. That extra boost diminishes with increasing latitude, so payloads destined for low-inclination orbits are usually launched from sites situated as close to the equator as possible. Polar-orbit payloads are launched southwards from Vandenberg. The first landmass encountered is Antarctica, so any failed shots fall harmlessly into the ocean.

With the advent of the space shuttle, the Cape Canaveral Air Force Station, from which low-inclination NASA and Air Force payloads have been launched for two decades, will be closed down. Shuttles will use NASA's Launch Complex 39, originally built for the Saturn V Apollo moon rockets.

On the West Coast, the Air Force is building new facilities and refurbishing others to accommodate the shuttle. Originally slated to be operational in October 1982, the Vandenberg facility is now scheduled to launch its first shuttle mission in October 1985.[25]

NASA launches sounding rockets and occasionally small earth-orbiting payloads from its Wallops Island, Virginia, facility.

U.S. Launch Vehicles

Over the last quarter century the United States has used four basic launch vehicles, configured with a bewildering variety of upper stages and strap-ons, to place military payloads into orbit. In order of increasing capability, these are the Scout, Delta, Atlas, and Titan. Starting in 1982, the shuttle also began carrying military cargoes and will become the principal vehicle for all U.S. launches by mid-decade.[26]

The Scout is a solid fuel, multi-stage booster capable of placing a 200-kilogram payload in low earth orbit. Its first stage was designed to be a submarine-launched ballistic missile, but the Navy chose a competitor. Its fourth stage, an Altair III, will be used as the second stage of the American air-launched ASAT. Scout has been used to orbit small scientific payloads and the Navy's Transit navigation satellites.

The Delta is a cryogenically fueled vehicle based on the Air Force Thor intermediate range ballistic missile. First used successfully as a launch vehicle in 1960, it could place barely ninety kilograms in low earth orbit. Eighteen years and twenty-five variations later, with added upper stages and nine solid strap-on boosters, payload capacity had grown to 1,815 kilograms. The increase was due largely to the strap-ons, which generate 83 percent of the first stage thrust. The Delta, in all its manifestations, has placed more than 300 payloads in orbit or on deep space trajectories.

Atlas was the United States' first ICBM. A cryogenic "stage and a half" design, it uses three first-stage motors sharing a common fuel tank, two of which drop away soon after liftoff. The basic vehicle was used to launch the first U.S. astronauts, in the Mercury program. With the Agena upper stage, Atlas could place about 2,700 kilograms in low earth orbit. (The Agena D, a restartable vehicle, was particularly useful with reconnaissance payloads needing considerable maneuvering capability.) With the Centaur hydrogen/oxygen upper stage, Atlas could boost a 3,850-kilogram payload into LEO, or 1,100 kilograms to escape velocity, and was used to launch Surveyor, Mariner, and Pioneer space probes to the moon, Mars, and Jupiter, respectively. Because Centaur is cryogenically fueled, it is not as useful as Agena for long-duration missions, but its high thrust (nearly twice the Agena) enables it to boost heavy satellites to GSO.

The Titan, another ICBM, was in its earliest version a cryogenically fueled missile. Test flights of a replacement using storable hypergolic propellants, the Titan II, began in 1962. It was chosen as the launch vehicle for the Gemini program, which tested human endurance of weightlessness as well as techniques of orbital rendezvous on behalf of the Apollo moon program. Also in 1962, Titan II was chosen as the Air Force's standard military launch vehicle. It was redesignated Titan III to distinguish it from the ICBM and configured several different ways.

The "A" variant was a testbed for a restartable vehicle called Transtage, which was capable of carrying several satellites at once and placing them in several different orbits. (Transtage was a precursor to the "post boost vehicles" used to send the multiple warheads atop modern ICBMs to their respective, widely dispersed targets.) The "B" variant operated only from Vandenberg Air Force Base and, with an Agena D as its third stage, could place about 3,000 kilograms in low earth orbit. Since all stages were hypergolic, the system was, theoretically, able to respond to short-notice launch requests. Between 1966 and 1979, fifty-one of fifty-eight launches probably involved reconnaissance satellites.

The Titan IIIC was a major innovation that used both the Transtage and two large strap-on solid boosters, each with three times the thrust of the basic first stage engines. Just as the "B" variant specialized in polar-orbit launches, the "C" specialized in placing one or more satellites (up to 1,200 kilograms total payload) in geosynchronous orbit, from Cape Canaveral. Substitute an Agena for Transtage and shift the launch back to Vandenberg, and you have a fourth, "D" variant used to launch very heavy (13,000 kilogram) reconnaissance payloads into polar orbit. Finally, NASA has used the Titan III to launch its heaviest interplanetary probes, the Viking Mars missions and the Voyager missions to Jupiter and Saturn.

The only missions requiring yet larger boosters were the Apollo moon program, Skylab, and the Apollo-Soyuz Test Project. Those missions used the Saturn. The largest configuration, Saturn V, stood 110 meters tall and could place 90,000 kilograms in low earth orbit or send 29,000 kilograms to the moon, which it did repeatedly from 1968 through 1972. An even larger booster, the Nova, was considered and dropped in 1963, when the United States decided against a direct earth-to-moon landing and return. For comparison purposes (and to keep in mind when reading about Soviet launch vehicles), the Nova could have put 160,000 kilograms in low orbit. It was not built because there was no mission for it.

All the boosters discussed so far are expendable, to be used only once and destroyed in the process. That is expensive. The hardware costs for a space launch may be ten to fifty times the costs of fuel consumed. So, theoretically, if a launch vehicle can be reused the cost of putting satellites into orbit should come down dramatically. That is one reason NASA started planning for the space shuttle in 1969, even as Neil Armstrong was taking a giant leap for mankind, courtesy of the Saturn V.

The shuttle is a major departure in launch vehicles. It can put up to 29,500 kilograms in low earth orbit and carry a crew of seven. The crew draws not more than three gravities acceleration on liftoff, and about half that on reentry (sufficiently low that, on reentry, crew members on mission seven could move about the cabin). Space shuttle main engines are designed to be used up to fifty-five times without overhaul, roughly seven-and-a-half hours of normal operations (compared to perhaps three minutes operating time for expendable first-stage engines).

The shuttle is limited to orbits of about 1,200 kilometers or less, and (without additional power supplies) to missions of not more than ten days duration. To reach higher orbits, payloads must have their own attached upper stages. Both the spinning solid upper stage (SSUS) and the inertial upper stage (IUS) substitute for the reusable space tug that was originally part of the planned shuttle system. The two versions of the privately developed SSUS will carry 1,100 and

2,000 kilograms to GSO, while the Air Force/NASA-funded IUS will take up to 2,300 kilograms to the same orbit.

Beginning in 1985, the shuttle will operate from Vandenberg as well as Kennedy, launching satellites into polar orbits. At about the same time, the U.S. Air Force plans to stop using the Titan and other expendable boosters.[27]

Soviet Launch Sites

The Soviet Union operates three space launch facilities. Tyuratam in Kazakhstan (latitude 45.6° north; longitude 63.4° east) is the largest, covering roughly 9,600 square kilometers. (The Soviets call it the Baykonur Cosmodrome, though it is 230 miles from the town of that name and is not on the same rail line. Its true map coordinates have yet to be made public officially by the Soviet government.) All manned flights, lunar and planetary missions, geosynchronous communications satellites, ocean surveillance and satellite interceptors, and most photoreconnaissance missions are launched from there, as are the heaviest Soviet launch vehicles. No American set foot at the site until the Apollo-Soyuz Test Project was flown in 1975, and then U.S. parties were flown to and from the site at night. Nearby is the "space city" of Leninsk, with a population estimated at 50,000.[28]

Plesetsk, the second Soviet launch site, in the Arctic on the rail line from Moscow to Archangel (latitude 62.8° north; longitude 40.1° east), started operations in 1966. The busiest launch site in the world, it is about one-third the size of Tyuratam. Unlike that complex, it is not officially acknowledged to exist. Plesetsk handles all polar-orbit missions, early warning satellites, and some photoreconnaissance. Because of its high latitude it is unsuited for equatorial-orbit missions but is ideally suited for missions using the stable 63-degree inclination. Such missions, of which there are many, can be launched due east to take advantage of what earth-rotation boost remains that far north.

The third Soviet site is near Kapustin Yar and is called Volgograd Station for its proximity to that city (the former Stalingrad). It is used to launch sounding rockets and smaller orbital payloads.

Soviet Launch Vehicles

The Soviet Union, like the United States, uses several launch vehicles derived from military missiles, plus a few heavy launchers built for the purpose. The launch vehicle names used here are not Soviet designators (those are not available in the open literature) but the system of designators developed by the late Charles S. Sheldon II, of the Congressional Research Service, a system widely used in nongovernmental publications on the Soviet space program.[29]

The Soviet government seems to believe that if something works, you should keep using it. The SS-6 ICBM, source of the 1957 missile gap hysteria and the launch vehicle for Sputnik I, is an ungainly vehicle, a slender core to which four large liquid-fueled strap-ons are attached, which gives the first stage a cone-like shape more than ten meters across at its base. That size and shape made the SS-6 difficult to base, and use of LOX meant that it could not be kept ready to launch. But those features do not detract from its use as a space launcher. Only four were deployed, briefly, as ICBMs, but through 1982 more than 950 had been launched into space, more than any other vehicle, American or Soviet. The launch frequency shows no signs of diminishing (fifty-eight in 1982). Designated "A" by Sheldon, it has been configured with a half-dozen different upper stages, and is the principal launcher for the Soviet manned space program and for most unmanned military and planetary missions. In its most capable configuration, it can place 7,500 kilograms in low earth orbit. It was this heavy lift capability that caused much of the early alarm in the West about the Soviet ICBM program.

The "B" and "C" vehicles were derived from the SS-4 and SS-5 intermediate range ballistic missiles first deployed by the Soviet Union in the early 1960s (including, nearly, in Cuba). The smaller B was phased out in 1977, but the C continues to launch communications and surveillance satellites from Plesetsk, and third-country satellites (like the Indian Bhaskara) from Kapustin Yar. The C uses hypergolic propellants, as do all current Soviet boosters except the venerable A's. Its orbital capability is roughly half that of the American Delta (900 kilograms to low orbit). The C is often used to place multiple small satellites in orbit.

The "D" vehicle, the largest, is also called the Proton, after the first series of satellites it launched in 1965. The Proton is used to launch all of the Soviet Union's geosynchronous satellites. It can heft 22,500 kilograms into low orbit or about 2,500 kilograms to geosynchronous orbit. The great difference is due in part to the energy-consuming, forty-five degree plane change needed to put a Soviet satellite into equatorial orbit, and in part to the Soviets' lack of a high energy, oxygen/hydrogen upper stage. Soviet boosters thus need much more initial thrust than do Western vehicles to put a given payload into geosynchronous orbit.

The final operational booster, designated "F," is based on the SS-9 ICBM, the largest of the 1960s generation of Soviet military missiles. Since it has been replaced in that mission by the newer SS-18, there are surplus boosters available for space applications, and F launches have increased. It has somewhat less payload capacity than the A vehicle but because of its storable propellants is more readily available for launch on short notice. It is the designated launcher for ocean surveillance, earth resources, and ASAT missions.

There is one launch vehicle whose debut has been expected for more than a decade. Thought to have been designed originally for the abortive Soviet moon landing program, the "G" design is speculated to have a lift capability to low orbit similar to the American Nova. The only payload of that mass (perhaps 180,000 kilograms) with any utility in low orbit would be a space station much larger than Salyut (see Chapter 3). Such a station is expected, and a "G" or booster of similar thrust would be needed if it were to be orbited in one piece.

The Soviet Union is also developing a reusable shuttle-like vehicle that might supply such a station. As of the end of 1983, three orbital tests had been conducted using a subscale model. The first two test flights ended in the Indian Ocean northwest of Australia. Photographs taken of the vehicle by Australian aircraft show a delta-winged lifting body. The third test flight ended in the Black Sea.[30]

A NEED FOR PEOPLE?

The Soviet space program did not invent the space station. That credit should probably go to Edward Everett Hale whose *Brick Moon,* published in 1870, featured a large, earth-orbiting aid to navigation. Space stations have also figured, off and on, in NASA planning since the mid-1960s (and appear, once again, to be moving to center stage). In countless pieces of speculative fiction written in the last few decades, large space stations have served as launch points for interplanetary voyages, astronomical observatories, or nodes at which characters' paths might cross. Computers have sometimes been the central characters in those stories, but by and large people have been necessary to keep things interesting. Unattended platforms were a rarity.

As the use of space has unfolded thus far, however, unattended operations have been the norm, in military, scientific, and commercial ventures alike. Indeed, in most of this book there will be relatively little reference to roles for people in space, and it is worth discussing briefly why that is the case.

Putting a crew into space incurs a number of costs with respect to launch, on-orbit life support, and reentry. Launch vehicles must be "man-rated," implying more stringent design criteria and slower accelerations for human safety and comfort. This affects launcher efficiency, since with slower accelerations more time is spent working against gravity. Bringing crews back requires heat shields for the vehicle, careful temperature regulation for the crew compartment, extra fuel for maneuvering, and ground logistics for vehicle retrieval. Safety is a paramount concern (losing a $50 million satellite is frustrating; losing an astronaut is tragic), and safety costs money.

In space, temperatures must be closely controlled for humans, as noted above, whereas sensors and other electronic instruments can operate over a much wider temperature range (from −55 to +125 degrees Celsius), and some sensors even prefer to operate near absolute zero (−273 degrees C.). Instruments are also easier to shield from ionizing radiation, most love to operate in vacuum, and few mind the absence of gravity. Humans, on the other hand, require substantial radiation shielding, a pressurized environment, and vigorous exercise to offset the affects of zero gravity on muscle strength and blood flow.

Finally, there is a psychological dimension not encountered by automata. Life in a small, confined environment, detached from the earth, bearing the continued possibility of life-threatening emergency, can produce psychological pressures that may interfere with normal human activity, especially the cognitive functions (judgment and reasoning).

The above considerations provide a feel for the complications involved in human space flight. Several of these factors only affect people; most affect both human and machine operation; but humans invariably are more sensitive. Given all of these costs, why would anyone want to put people into space? One reason is national pride, which played no small part in Soviet efforts to make Gagarin the first man to orbit the earth and American efforts to place the first man on the moon. Because sending humans into space is so difficult, doing so demonstrates technical prowess and, by extension, a nation's superiority. However, with these and other space "firsts" firmly claimed on behalf of national pride, we are relatively free to contemplate what useful role humans might actually play in space.

Humans are particularly adept at pattern recognition, making judgments, reasoning inductively, and improvising. They have a large memory capacity from which pertinent information can be drawn and applied to the task at hand. Aside from such cerebral functions, humans possess considerable manual dexterity, which allows them to perform complex manipulations with tools. On the other hand, humans are not as well equipped to sense physical quantities such as temperature, pressure, and levels of electromagnetic radiation. For routine tasks and calculations, machines are more reliable, faster, and more powerful. Computers, in particular, offer large, quick-access memory and a better erase capability (so that new information can be reliably stored). Consequently, machines can be made more sensitive to physical inputs and can respond to commands more quickly than humans.

Since sensing electromagnetic radiation is something humans do very poorly with the exception of a small band of frequencies sensed as visible light (and even then not as well as modern optical instruments), they can be excluded immediately from most current military space activities (the reconnaissance, surveillance, and early warning activities discussed, along with other missions, in Chapters 2 and 3). To capitalize on human cognitive functions it will probably always be more advantageous to transmit information to ground operators, rather than have humans stationed in space.

In terms of future applications—space based weapons, for example—the potential role for humans is not obvious. Most modern weapons rely on sensors for target detection and tracking, and for warhead guidance. In many cases, humans intervene only in the decision to fire. For space-based weapon platforms, target acquisition and guidance would certainly be performed by automatic sensors. Weapon delivery could also be accomplished without "hands on" human assistance. Space flight bears no resemblance to flying an airplane, since maneuvering between orbits is restricted and predictable. Even an order to fire would best be accomplished from the ground via secure communication links, especially considering the amount of time such weapons would remain essentially idle in orbit (from months to forever) before being called into use. Supporting crews for this type of mission and for that length of time would involve tremendous costs, sheer boredom being added to the list discussed above.

The one human faculty that might be important for future space missions is manual dexterity. The ability to carry out complex operations with tools means that construction and repair of large space structures might require human presence. (Whether large space structures are themselves necessary or desirable is a separate question.) Expensive satellites that have failed might be repaired in space, if the repair job is relatively simple. For large or complex tasks, the size of the human workforce would be large, because a strategy of on-orbit repair would demand a wide range of skills and a wide variety of diagnostic support equipment and spare parts. Such skills, equipment, and parts might be consolidated at a large space platform, but the question remains whether it would be more cost-effective to maintain such an endeavor in orbit, resupplying it by shuttle, or to use the shuttle and small orbital transfer vehicles to retrieve crippled spacecraft and bring them back to earth, where repair crews could breathe without support, eat at McDonald's, go home at night and to the beach on weekends. There's not much romance in that, but there is a certain logic.

NOTES

1. Arnold Toynbee, *A Study of History,* volume one of the abridgement by D.C. Somervell (London: Oxford University Press, 1946), pp. 251, 374.
2. Ibid., p. 375.
3. Joseph Swain and William Armstrong, *Peoples of the Ancient World* (New York: Harper & Row, 1959), p. 306; and Carl Sagan, *Cosmos,* (New York: Random House, 1980), pp. 188–89.
4. Thomas S. Kuhn, *The Copernican Revolution* (Cambridge, Mass.: Harvard University Press, 1957), pp. 185–99; and Willy Ley, *Rockets, Missiles and Space Travel* (New York: Viking, 1968), pp. 10–11.
5. Samuel Glasstone, *Sourcebook on the Space Sciences* (Princeton: C. Van Nostrand, 1965), pp. 457–59.
6. Ibid., pp. 524, 543–45; and National Research Council, *Outlook for Science and Technology* (Washington: National Academy of Sciences, 1982), p. 344.
7. Ibid., p. 363; and Glasstone, pp. 897–99.
8. See James F. Parker, Jr. and Vita R. West, managing eds., *Bioastronautics Data Book,* 2d ed. Sponsored by the Scientific and Technical

Information Office, National Aeronautics and Space Administration. SP-3006. (Washington, D.C.: GPO, 1973), Chapter 3.

9. Ley, pp. 9, 17.
10. Wernher von Braun and Frederick Ordway III, *The Rockets' Red Glare* (Garden City, N.J.: Doubleday and Co., 1976), pp. 121, 124; and Glasstone, p. 22.
11. Ibid., pp. 18–22, 38.
12. P.G. Hill and C.R. Peterson, *Mechanics and Thermodynamics of Propulsion* (Reading, Mass.: Addison-Wesley, 1965).
13. U.S. Congress, House Committee on Science and Technology, *U.S. Civilian Space Programs, 1958–78, Volume I,* Committee Print, Serial D, January 1981, pp. 221–22.
14. Von Braun and Ordway, pp. 31, 34, 46.
15. Ibid., pp. 71, 77; and Bernard Brodie and Fawn M. Brodie, *From Crossbow to H-Bomb* (Bloomington: Indiana Univ. Press, 1973), p. 227.
16. Ibid., p. 134; Glasstone, pp. 23, 35.
17. Von Braun and Ordway, p. 138.
18. Glasstone, p. 26.
19. Ibid., pp. 35–36; Ley, pp. 313–15.
20. Glasstone, p. 34.
21. Ley, pp. 304–9.
22. Ibid., p. 313.
23. Ibid., p. 317.
24. Glasstone, pp. 33–76.
25. U.S. Congress, House Committee on Appropriations, *Military Construction Appropriations for 1983, Part 4, Hearings before the Subcommittee on Military Construction,* 97th Cong., 2d sess., 10 March 1982, p. 429.
26. U.S. launch vehicle data has been drawn from *U.S. Civilian Space Programs,* and from Marcia S. Smith, "Space Shuttle Issue Brief," Congressional Research Service, Washington, D.C., 16 February 1983.
27. *Aviation Week & Space Technology* (6 June 1983): 89.
28. Information on Soviet launch sites and vehicles has been drawn from U.S. Congress, Senate Committee on Commerce, Science and Transportation, *Soviet Space Programs, 1976–1980, Part I,* Committee Print, December 1982.
29. The system incorporates features first used in the *TRW Space Log.*
30. Craig Covault, "Soviets Orbit Shuttle Vehicle," *Aviation Week and Space Technology* (14 June 1982): 18–19; and Thomas O'Toole, "Soviets Orbit Space-Shuttle Prototype," *Washington Post,* 28 December 1983.

REFERENCES

Aviation Week and Space Technology (6 June 1983): 89.

Brodie, Bernard M., and Fawn M. Brodie. *From Crossbow to H-Bomb.* Bloomington: University of Indiana Press, 1973.

Covault, Craig. "Soviets Orbit Shuttle Vehicle," *Aviation Week and Space Technology* (14 June 1982): 18–19.

Glasstone, Samuel. *Sourcebook on the Space Sciences.* Sponsored by the National Aeronautics and Space Administration. Princeton: C. Van Nostrand, 1965.

Hill, P.G., and C.R. Peterson. *Mechanics and Thermodynamics of Propulsion.* Reading, Mass.: Addison-Wesley, 1965.

Kuhn, Thomas S. *The Copernican Revolution.* Cambridge, Mass.: Harvard University Press, 1957.

National Research Council. *Outlook for Science and Technology.* Washington National Academy of Sciences, 1982.

O'Toole, Thomas. "Soviets Orbit Space-Shuttle Prototype." *Washington Post,* 28 December 1983.

Parker, James F., Jr., and Vita R. West, managing eds. *Bioastronautics Data Book,* 2d ed. Sponsored by the National Aeronautics and Space Administration. NASA SP-3006. Washington, D.C.: GPO, 1973.

Toynbee, Arnold. *A Study of History.* Volume one of the abridgement by D.C. Somervell. London: Oxford University Press, 1946.

U.S. Congress. Congressional Research Service. "Space Shuttle Issue Brief." by Marcia S. Smith, 16 February 1983.

———. House. Committee on Appropriations. *Military Construction Appropriations for 1983, Part 4, Hearings Before the Subcommittee on Military Construction.* 97th Cong., 2d sess., March 1982.

———. Committee on Science and Technology, *U.S. Civilian Space Programs, 1958–1978, Volume I.* Committee Print. Serial D. January 1981.

———. Senate. Committee on Commerce, Science, and Transportation. *Soviet Space Programs, 1976–1980, Part I.* Committee Print, Dec. 1982.

Von Braun, Wernher, and Frederick W. Ordway III. *The Rockets' Red Glare.* Garden City, N.J.: Doubleday and Co., 1976.

2 SPACE AND U.S. NATIONAL SECURITY

Paul B. Stares

With the demise of U-2 aerial reconnaissance flights over the Soviet Union in May 1960, the United States immediately became dependent on reconnaissance satellites as the primary source of strategic intelligence on its cold war adversary.[1] So began a vital relationship of dependency on military satellites that has been progressively reinforced as the use of space has played an ever increasing role in performing or supporting other military operations.[2] This chapter provides an overview of the evolution of the U.S. military space program to illustrate the importance of outer space to U.S. national security. Particular attention will be paid to the formulation of U.S. space policy, the organization and management of the space program, the development of key satellite programs, and salient operational trends.[3]

EVOLUTION OF U.S. MILITARY SPACE POLICY

The United States did not have a military space program of any real substance until the Air Force was given permission to begin development of a reconnaissance satellite in 1954. However, feasibility studies of the military utility of artificial satellites had begun a decade earlier at the end of World War II, the most significant being the Rand studies sponsored by the Army Air Force. These were remarkably prescient in forecasting not only the likely psychological and political repercussions that the first satellites would have but also the range of military uses to which satellites would eventually be put. Despite this early appraisal of the benefits of satellites, a variety of factors—principally, political indifference, technological constraints, military conservatism, the austerity of post-war defense budgets and inter-service rivalry—hindered the development of a number of promising satellite proposals.

By the spring of 1954 an equally influential confluence of factors began to undermine the prevailing inertia and promote satellite development. The recommendation of the Strategic Missile Evaluation Committee in February 1954 to accelerate ballistic missile development also meant that the necessary technology to launch reasonably heavy satellites would be available for a space program. More significantly, the requirement for reliable strategic intelligence that was one of the major recommendations of another influential committee—the Technologies Capabilities Panel—in March 1954, encouraged the CIA and the Air Force to consider seriously the development of reconnaissance satellites as a long term alternative to aircraft. Its report coincided with the completion of a Rand study, "Project Feedback"—which also recommended that the "Air Force undertake at the earliest possible date, completion and use of an efficient satellite reconnaissance vehicle as a matter of vital strategic interest to the United States." By July 1954, approval had been given to start development of an "Advanced Reconnaissance System" (later known as WS-117L) with Lockheed eventually being awarded the prime contract in 1956. This became the basis of both the Air Force's and CIA's reconnaissance satellite programs—SAMOS (Space and Missile Observation System) and Corona, respectively.

Although a small U.S. military space research and development program was underway by 1957, the launch of Sputnik I in October provided the most important stimulus to satellite development. The widespread concern that the United States was somehow "behind" the Soviet Union produced an irresistible pressure for a rapid expansion of the U.S. space program. By the end of the Eisenhower administration, the foundations of each of the major military space programs had been laid. Similarly, between October 1957 and October 1963 the policy guidelines that have determined the subsequent U.S. exploitation of space were also formulated. Successive administrations have reaffirmed these guidelines with relatively few diversions or contradictions.

The initial concern of U.S. policymakers after Sputnik was to dispel the fear that the Soviet achievement was a harbinger of larger threats from space and of a dramatic shift in the strategic balance on Earth. The Eisenhower administration's confidence in the relative position of the United States was based almost solely on the evidence supplied by the highly secretive U-2 flights over the Soviet Union. Although these flights were still in their heyday, U.S. planners knew that persistent Soviet attempts to shoot down the U-2s would one day prove successful and thus bring to an end this major source of strategic intelligence. While this could be replaced by the use of reconnaissance satellites, it was by no means clear this would go unchallenged by the Soviet Union. As a result the necessity of safeguarding the principle and practice of reconnaissance from space—especially after U-2 flights over the Soviet Union were suspended in 1960—became the dominant objective in U.S. space policy.

This objective posed a considerable challenge to U.S. planners. First, the Soviet Union's understandable hostility towards aerial overflight of its territory made it highly likely that it would not let U.S. reconnaissance from space go unopposed—by either diplomatic or military means (see Chapter 8). In fact, the Soviet Union had every incentive to oppose the U.S. military use of space. Due to the closed nature of Soviet society, the United States would clearly be the beneficiary of satellite reconnaissance. It would benefit as well from access to long distance satellite communications with its far-flung allies and forces.

Second, the inherent fragility and predictable path of satellites made them particularly vulnerable to direct interference from a determined attacker. This could not be alleviated in any meaningful way either by hardening the satellites or by increasing their maneuverability as payload limitations in the early 1960s placed severe constraints on all but essential items. Third, because it was likely to become the more space dependent power, the United States could not adequately deter Soviet interference with U.S. satellites by threat of reciprocal action.

Faced with a problem that the traditional methods of enhancing security (defense or deterrence) could not adequately solve, U.S. policy makers pursued the only real alternative—namely, international sanction of military space activities. In this respect the United States had the advantage of being able to influence the development of international space law, which was still essentially a *tabula rasa*. In support of this goal, U.S. policy towards the military use of space followed four basic guidelines, albeit somewhat inconsistently in the initial period.

First, the United States promoted the objective that space was to be used for *peaceful*—that is, *nonaggressive*—purposes. This was a subtle but important shift from the earlier stated goal that space should be used for nonmilitary purposes.[4] As a corollary, the United States proposed that space would not be subject to national jurisdiction or sovereignty as it applied to air space and that states should have the right to full and unhindered passage through space. This goal was officially stated in the two space policy directives of the Eisenhower administration.[5] It was also reiterated in the public statements of the Kennedy administration, especially after the Soviet Union launched its long awaited diplomatic offensive in the United Nations to make "espionage satellites" illegal.[6] Although the ensuing debate over the legitimacy of military satellites subsided after the Soviet Union withdrew its objections in October 1963, recent concern over the possibility of Soviet antisatellite activities have led to a restatement of the legal status of satellite objects both in the public statements of President Carter (in a press release of Presidential Directive 37 in June 1978) and in the Reagan administration's National Security Decision Directive 42 on space policy announced on 4 July 1982.[7]

Second, the United States reduced the public profile of its military space activities to avoid contradicting publicly its commitment to the peaceful use of space, and at the same time to minimize potential political and military opposition from abroad. This process began with the Eisenhower administration's attempts to conceal the true purpose of the Corona program by giving it the public name Discoverer and a cover description for its mission (biomedical research and development). It was finally completed in March 1962 by a formal Department of Defense directive that essentially stated that the U.S. satellite reconnaissance program and any other sensitive operations in space would no longer be publicly acknowledged. While the constraints on public references to the other parts of the military space program have gradually been relaxed, this directive has remained strictly in force with regard to U.S. satellite reconnaissance activities. It was relaxed only marginally when President Carter acknowledged in 1978 that the United States operated satellites with this purpose.

Third, the United States would show restraint in the development of weapon systems for use in or from space. American policymakers recognized from an early date that there was little military incentive for the United States to develop "space weapons." Orbital bombardment systems, for example, did not offer the same capability as ballistic missiles due to their limited operational flexibility, control, and accuracy. Furthermore, despite earlier fears, there was never an unambiguous Soviet "space threat" to justify an extensive antisatellite (ASAT) program. More importantly, it was felt by both the White House and the Defense Department that a vigorous U.S. ASAT program would encourage the Soviets to develop similar systems of their own that, given the growing U.S. dependence on military satellites, would do more harm to American than to Soviet interests. The only apparent exceptions to this guideline were the two land-based ASAT systems that were deployed on Johnston Island and Kwajalein Atoll in 1963–64. These, however, can be categorized more correctly as orbital bomb defense systems deployed as insurance against some future Soviet attempt to blackmail the United States with the threat of bombardment from space. Although one of the U.S. ASAT systems remained nominally operational until 1975, it was recognized that the use of its nuclear warhead would pose an equal risk to U.S. satellites in the vicinity of the explosion while its fixed site limited its response time and the number of targets it could attack. Thus, it had a questionable utility in all but extreme scenarios. The services also undertook considerable research into a variety of space weapon concepts, but this research did not go beyond feasibility studies and was intended to maintain a pool of expertise in case U.S. policy demanded further development.

This policy only begin to change in the late 1970s with President Ford's decision in the last days of his administration to pursue ASAT capability. The decision was prompted in part by the February 1976 resumption of Soviet ASAT tests (after a four-and-a-half-year hiatus) and the perceived indirect threat to U.S. terrestrial forces from certain Soviet reconnaissance satellites. The Carter administration endorsed Ford's decision but as part of a "twin track" policy that included an attempt to reach an ASAT arms control agreement with the Soviet Union. The endorsement was designed to provide bargaining leverage with the Soviet Union during the negotiations and insurance if they proved unsuccessful.

Fourth, in support of the first three guidelines, the United States encouraged international agreements and treaties that codified the legitimacy of peaceful military activities in space and prohibited those activities deemed inimical or superfluous to military requirements. Since 1962, the United States has entered into a variety of agreements of both direct and indirect relevance to military activities in space. These include the Partial Test Ban Treaty of July 1963 that prohibited, inter alia, nuclear explosions in space; the October 1963 UN Resolution banning the deployment of weapons of mass destruction in space; the December 1963 Declaration of Legal Principles Governing Activities in the Exploration and Use of Outer Space (without the original Soviet objection to "espionage satellites"); the 1967 Outer Space Treaty, which codified the nonbinding 1963 UN resolution banning weapons of mass destruction in space; the clause in the 1972 SALT I agreements (repeated in subsequent arms control agreements) prohibiting interference with "national technical means of verification," generally considered to include reconnaissance satellites; as well as the ABM Treaty prohibition on the development, testing or deployment of space-based ballistic missile defenses (for further discussion, see Chapter 7). In addition, as a result of President Carter's initiative, the United States and the Soviet Union began negotiating in 1978, albeit unsuccessfully, an agreement to limit the further development of antisatellite weapons.

Three sets of talks were held in Helsinki 8–16 June; in Bern from 23 January to 19 February 1979; and finally in Vienna from 23 April to 17 June 1979. Although the United States hoped that a comprehensive agreement could be reached that would ban the testing and deployment of antisatellite weapons, progress seems only to have been made towards defining a "non-use" agreement with additional "rules of the road" for activities in space. The most significant obstacles to an agreement were reported to be the Soviet insistence that the Space Shuttle be included in the negotiations and that "third party" satellites be excluded. Both conditions were unacceptable to the United States. The rupture in East-West relations following the Soviet invasion of Afghanistan in December 1979 put paid to any further rounds of talks.

Given the early concern and uncertainty over Soviet intentions towards space, the subsequent militarization of this medium represents a qualified success for U.S. policy. The military uses of space that the United States most coveted have not been hindered by legal constraints or, until comparatively recently, by the threat of direct Soviet interference. Although this was arguably the result of the Soviet Union's growing military use of space, U.S. policy was a major factor in the modus vivendi that was reached with the Soviet

Union. This has allowed both to conduct their military operations in a mutually beneficial way.

THE ORGANIZATION OF THE U.S. SPACE PROGRAM

With the widespread public and congressional clamor for a reinvigorated national space program after the launch of Sputnik, Eisenhower and his advisers were immediately faced with the problem of organizing the U.S. effort. As in the case of space policy formulation, the organization and management decisions of the Eisenhower and Kennedy administrations have had a profound impact on the subsequent development of the space program.

Eisenhower's task had already become complicated by the decision in May 1955 to develop a scientific satellite (later called Vanguard) with the strict criterion that any military involvement in its development would be kept to a minimum.[9] Although this was designed to avoid hindering the ballistic missile effort and enhance the "peaceful" image of U.S. space activities, the precedent for a bifurcated space program had been established. However, with the embarrassing early sequence of Vanguard test failures, it was clear that a "civil" space program would need—certainly in its early stages—substantial help from the military. Furthermore, in the aftermath of Sputnik, each of the services pursued their own space programs (some of which, like WS-117L, had begun before 1957) to take advantage of the likely budgetary payoffs from the crisis. Thus Eisenhower was confronted with the problem of reconciling both the civil and military elements of the national space program as well as controlling the proliferating number of space projects.

In an effort to avoid duplication and rivalry among the competing services, the Advanced Research Projects Agency (ARPA) was established in November 1957 to concentrate authority for all space projects within the Department of Defense. Eisenhower hoped that civil/scientific space research would be coordinated by the DoD, but pressure from Congress and the President's Science Advisory Committee led to creation of a separate civilian agency—the National Aeronautics and Space Administration (NASA)—in April 1958.[9]

These initiatives proved to be only partially successful. ARPA's role in coordinating the military space program was short lived as duplication and rivalry, if anything, increased. As a result, the responsibility for the development of military space projects was transferred back to the services in September 1959. The Civil-Military Liaison Committee that had been created to coordinate NASA and DoD space activities was also unsuccessful and subsequently replaced by the Aeronautics and Astronautics Coordinating Board in July 1960. This continues to function today.

Concern over the poor coordination and progress with the highly classified satellite reconnaissance projects also led to the creation in August 1960 of a separate organization—the National Reconnaissance Office (NRO)—to oversee operations. Although the NRO was located within the Department of the Air Force, it coordinates the intelligence gathering activities of the CIA, NSA, the U.S. Air Force and the U.S. Navy. While the Air Force has primary responsibility for the launch and station keeping of the satellite vehicles, the CIA has a major role in the design and development of the payload.[10] With the development of electronic intelligence gathering, the National Security Agency (NSA) also took a more active role in U.S. covert space programs. In addition to aiding the development of some ELINT gathering equipment aboard U.S. satellites, the NSA also formed the Defense Special Missile and Astronautics Center (DEFSMAC) in September 1966 that apparently coordinates the targeting of items of interest to the intelligence agencies as well as their eventual analysis.[11]

Despite Eisenhower's later adjustments, the organization of the military space programs was severely criticized by a study group led by Jerome Wiesner for President-elect Kennedy. As John Logsdon recounts, its final report "deplored the tendency of each military service to create an independent space program and it called for the establishment of a single responsibility for a space program among the military services."[12] The new Secretary of Defense, Robert McNamara, immediately authorized a review of the organization for military space research and development, which resulted in Defense Directive 5160-32 "Development of Space Systems," signed on 6 March 1961. This permitted each of the services to conduct preliminary research within guidelines prescribed by the Director of Defense Research and Engineering (DDR&E). Thereafter, further development would be the prerogative of the Air Force unless the initiating service could make a very strong case for holding on to it.

The net result was a hollow victory for the Air Force, which had been campaigning vociferously for the space mission on the grounds that space was merely an extension of the air medium. Rather than gaining sole responsibility for military space activities, the Air Force in practice continued to share responsibility. This directive, which has remained in force for more than two decades with only some minor modifications, has had a profound effect on Air Force and other service attitudes to space. Without an operational monopoly the Air Force agencies responsible for space have remained the poor relations within that service's hierarchy. This was further reinforced by the Air Force's failure to gain a manned military space program of its own, despite coming tantalizingly close to this goal with such programs as Dyna-Soar, Blue Gemini and the Manned Orbiting Laboratory; all were eventually cancelled in the 1960s, which further demoralized and alienated supporters of space systems within the Air Force.

The inherent character of the benefits provided by satellites has also had a major impact on service attitudes towards space. The main benefits in such areas as intelligence gathering, communication, navigation, and weather forecasting are not service-specific. With the divided responsibility for space development, the services (and intelligence agencies) have in many instances either unnecessarily duplicated certain satellites or been reluctant to propose new systems that might eventually entail the use of their own funds to help other services. The initial development of the NAVSTAR GPS navigation satellite is a prime example of the problem of gaining support for a system that is clearly beneficial to all the potential users.[13]

This phenomenon was particularly evident in periods of fiscal constraint. After the boom years of the space budget in the late 1950s and early 1960s, funding for space plummeted in the late 1960s and early 1970s. Figure 2–1 graphs DoD and NASA space expenditures (budget authority) in billions of current dollars to illustrate these budgetary fluctuations.

Since the late 1970s the budgetary trend has been upwards, and consequently service interest has been more supportive of space activities. This was reflected by Air Force's decision to upgrade the status of its space agencies within the service, culminating in the creation of a separate Space Command in September 1982.[14]

Figure 2–1. DoD and NASA Space Expenditure (Budget Authority), Fiscal Years 1959–84 (est.).[a]

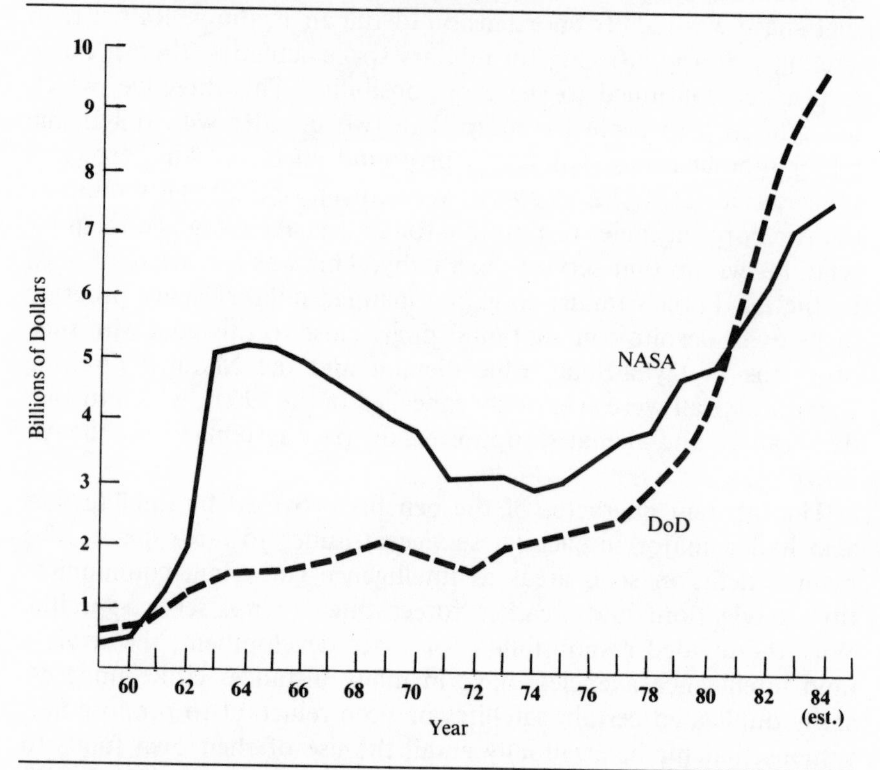

[a] DoD figures do not include NRO or NSA space national security programs or defense-related research carried out by NASA.

Sources: *Aeronautics and Astronautics Report of the President, 1980,* p. 102; and Sarah Sewell, "Militarizing the Last Frontier," *Defense Monitor* 12:5 (1983).

EVOLUTION OF U.S. MILITARY SPACE PROGRAMS

The U.S. military space program has remained essentially an unmanned enterprise. Despite numerous studies examining the utility of manned military operations in space, all have reached the same general conclusion that there was little that could not be achieved by automated systems. The engineering problems and additional costs associated with sustaining manned activities in space made it doubly unattractive. The increasing sophistication of unmanned space platforms has underlined this fact still further. Only the planned DoD operations using the space shuttle offer the prospect of a U.S. manned military presence in space.

The evolution of the military space program can be arbitrarily divided into six functional areas of development: reconnaissance, surveillance, communication, navigation, meteorology, and geodesy. It is an arbitrary division as the trend towards multifunctional or "integrated" space systems has blurred the distinction between many of these categories. Before analyzing some of the more salient operational trends that cut across these separate categories, it is useful to begin by reviewing the most significant milestones in the development of these programs.

Reconnaissance

The U.S. reconnaissance satellite program can also be broadly divided into three separate but related categories: photographic, electronic intelligence gathering, and ocean observation.

Photographic satellites are used for a variety of purposes, including strategic intelligence gathering, tactical reconnaissance, arms control verification, crisis monitoring, and post-attack assessments. The original SAMOS and Corona programs became the first operational systems in what would subsequently be known as the Keyhole (KH) series; SAMOS becoming KH-1 and Corona KH-4.[15] They were distinguished by their data transmission techniques and their respective intelligence missions. The CIA's KH-4 returned its exposed film by recoverable capsule. This proved to be more difficult than at first thought as it was not until the thirteenth attempt, 10 August 1960, that this technique was successfully demonstrated. With the perfection of these operations the Corona satellites were progressively used for "close-look" photographic missions. The close-look/film return satellites have since then undergone second and third generation changes that have improved the resolution of the main camera, prolonged their orbital life expectancy, and possibly provided additional sensors. The first of the second generation close-look satellites (KH-6) was launched on 12 July 1963 by Atlas Agena D booster. These second generation satellites remained in orbit an average of four to five days. The initial third generation close-look satellite (KH-8) was launched on 29 July 1966 by a Titan IIIB Agena D booster.

The operational lifetime of these satellites, which were still being launched in 1983, has been progressively extended to over a hundred days.

In contrast, the Air Force SAMOS relied on the more ambitious technique of transmitting its film by radio links to ground stations. Like the Corona program, SAMOS was dogged by teething troubles, and it was not until SAMOS 6, launched on 7 March 1962, that the first fully successful mission was completed.[16] As the resolution of its TV cameras was not of the caliber of the Corona system, SAMOS was used principally for search and find missions (i.e., area surveillance). Resolution was improved in a further two generations of SAMOS-type satellites—the KH-5 and the KH-7—which took advantage of the extra lift capacity of successive U.S. launchers.[17]

By the beginning of the 1970s the operational responsibility for these two types of intelligence missions appears to have been reversed with the fourth generation of photographic reconnaissance satellites. In addition to the KH-8 operations mentioned above, the Air Force began operating what is commonly referred to as the "Big Bird" satellite (designated KH-9) from June 1971. While the Big Bird's camera system is of high resolution and returns its film by recoverable capsules, it is generally considered to be used for search and find missions. The CIA's KH-11 satellite, however, first launched on 19 December 1976, and which reportedly deploys a multifaceted array of sensors such as multispectral scanners (and possibly a synthetic aperture or side looking radar) transmits its information digitally in real time.[18]

The electronic intelligence (ELINT) or "ferret" satellites complement the photoreconnaissance mission in a number of important ways. These satellites are used for communications intelligence gathering that can give indications of the size, deployment, and readiness of military forces; for plotting the location and frequencies of air defense and missile early warning radars for direct targeting and electronic countermeasures; and for ballistic missile test monitoring for arms control verification.

Both the Air Force and the Navy began developing their own ELINT satellites in the early 1960s.[19] Similarly the CIA/NSA began their own ferret satellite programs that "hitchhiked" or rode "piggyback" with the Corona satellites. Subsequently, both the Air Force and the Navy have used this technique to launch their own ELINT payloads into space.[20] Although this practice continues, the advent of a new generation of ELINT satellites in the early 1970s represented a significant departure in electronic reconnaissance operations. These highly classified satellites operate at geosynchronous altitudes rather than low earth orbit. The first ELINT systems deployed in this orbit appear to have been the CIA's Rhyolite satellites designed to collect the telemetry from Soviet ballistic missile tests. Although the first launch in the Rhyolite series is generally considered to have been on 6 March 1973, it is highly likely that full or component testing began somewhat earlier under cover of the ballistic missile early warning satellite program (Code 949) and later the Defense Support Program (DSP or Code 647), which also has an early warning function. Considerable circumstantial evidence suggests that this was as early as 1970.[21] A more advanced follow-on system reportedly called "Argus" (though almost certainly referred to by a new code name) is also believed to have been tested under the cover of the Defense Support Program.[22] A similarly deployed NSA/Air Force satellite system under the code name "Chalet" is also reported to have been developed for general communications intelligence gathering.[23]

The U.S. Navy satellite ocean reconnaissance program has the primary objective of locating, classifying, and targeting enemy shipping. While it is likely that some of the Navy's ELINT satellites launched during the 1960s and early 1970s acted in this role, the first dedicated ocean reconnaissance program appears to have been the White Cloud system that was first launched on 30 April 1976. Once in space, three subsatellites spread out from the parent spacecraft in a 700-nautical-mile circular polar orbit to monitor a wider area on each pass. By use of millimeter-wave radio receivers and possibly some infrared detectors, surface shipping can be located.

The Navy also began developing a radar-based ocean reconnaissance system under the code name "Clipper Bow." Although Congress withdrew funding for this program in 1979, the Navy is currently funding a similar system as part of its Integrated Tactical Surveillance system (ITSS) and also pursuing the possibility of a cooperative venture with the Air Force. The potential of this technique was demonstrated with considerable success by NASA's Seasat program, which used side-looking radar for oceanographic purposes. This is another area of activity where the use of space offers considerable advantages to the military. Such information as the height of waves, the strength and direction of ocean currents and surface winds; sea and undersea temperatures; sea salinity; and coastal features that can all be obtained by space-based oceanographic sensors are important for naval activities generally and potentially vital to future antisubmarine warfare operations (see Chapter 5).

Surveillance

Reconnaissance and surveillance satellites are distinguished here on the basis of the difference between searching and watching, respectively. The two basic types of surveillance satellites are for nuclear explosion detection and ballistic missile early warning.

The first satellite with the explicit purpose of increasing strategic warning time over that offered by ground-based radars was the MIDAS program, which also had its origins in the early WS-117L studies. Unfortunately, its infrared sensors, designed to detect the hot exhaust plumes of ballistic missiles in their boost phase, were not sufficiently sensitive or reliable, and this program was cut back in late 1961. By the mid-1960s work had begun on a more refined early warning system (Code 979) that was launched into geostationary orbit for the first time on 6 August 1968. This was augmented (and later superseded) in 1970 by the first of the integrated early warning satellites under the Defense Support Program (Code 647), which utilizes a high resolution telescope with infrared sensors for the early warning role and some nuclear explosion detection sensors.

The idea of a space-based worldwide nuclear explosion detection system gained considerable support during the negotiations leading to the Partial Test Ban Treaty of 1963. It was also recognized that such a system would contribute to post-attack assessments during wartime. The first satellites of this type were the Vela Hotel series that began operations in October 1963. These were launched in pairs into high-altitude circular orbits for long orbital life and worldwide coverage.

Although the last two Vela Hotel satellites were launched in 1970, they are believed still to be operational today. In the future, their function will be performed principally by the NAVSTAR GPS satellites, which also carry sensors known as Integrated Operational Nuclear Detection Systems (IONDS).

Communications

The benefits of using satellites for communicating rapidly and reliably over long distances were recognized from an early date and these satellites were, therefore, among the first to be developed. During the late 1950s a variety of projects were funded by ARPA and the services but with mixed results.[24] It was not until the Initial Defense Communication Satellite Program (IDCSP) of the mid-1960s that the potential of communication satellites began to be realized, especially after their deployment was moved to the geostationary orbit. The IDCSP—renamed Defense Satellite Communication System (DSCS)—has subsequently undergone various improvements in channel capacity, anti-jam (AJ) facilities, and nuclear effects hardening with the DSCS II and, most recently, DSCS III programs. At the same time as these DoD-wide satellites were under development, the Air Force and the Navy argued that their specific communication requirements warranted their having their own satellite systems. As a result, the Navy developed the FLTSATCOM satellite system and the Air Force developed the Satellite Data System (SDS) in addition to deploying its own transponders, known as the AFSATCOM system, on a variety of other host satellites.

Navigation

The U.S. Navy proved to be most interested in developing satellites for navigation purposes as a means of updating the inertial navigation systems and missile targeting data aboard its Polaris submarines. The resultant Transit satellite program, which exploits the Doppler shift effect in the signal tone of passing radio signal transmissions, became operational in 1964. The service provided by Transit was later extended to surface shipping and is still available today for both military and civilian users. Its successor, the aforementioned NAVSTAR GPS, is being developed by the Air Force and will be available for use by all three services.[25] The first in a planned constellation of eighteen satellites was launched on 22 February 1978. Such is the accuracy of the information from NAVSTAR that it also offers a revolutionary improvement in weapons-systems targeting and guidance.

Meteorology

Meteorological satellites are probably the most underrated of the range of satellites discussed here. Weather prediction and real-time weather surveillance is not only important for general purpose military operations but crucial to strategic war planning. For example, timely information on cloud cover over part of the Soviet Union allows the more efficient use of U.S. reconnaissance satellites. Furthermore, the weather over a target area is likely to be a critical factor in the final accuracy of ballistic missile reentry vehicles, necessitating accurate meteorological information for preattack planning. Although NASA began developing a weather satellite for civilian use known as TIROS, its resolution was not of the quality needed to meet military requirements. Thus the Air Force began developing its own system, which, after some early failures, became operational at the beginning of 1962. On average, two or three weather satellites have been launched every year since then with successive improvements to their accuracy and data handling. The latest series is the Block 5D, first launched in September 1976.

Geodesy

Geodetic satellites are similarly underrated in their overall contribution to military operations. This category of satellites provides vital information on the size and shape of the earth's surface and its shifting gravitational fields, essential to the accurate mapping of the earth's surface. Needless to say, such information is vital for intercontinental ballistic missile targeting. The U.S. Navy began development of geodetic satellites with its ANNA program, which ran from 1964 to 1969. By 1972, responsibility for geodetic surveying was formally shifted to the Defense Mapping Agency. Subsequent geodetic mapping of the earth has been conducted by using NASA satellites such as the GEOS and LAGEOS systems.

OPERATIONAL TRENDS

Various factors have influenced the operational characteristics of the American military space program. Foremost among these has been the growth in lift capacity through the development of larger boosters. Working from a different direction, the increasing sophistication and miniaturization of electronic equipment has also permitted weight saving efficiencies in payload design. The net result has been the deployment of ever more capable payloads in orbit. This has in turn influenced the number of launches required each year, the choice of orbits that can be utilized, and the operational lifetimes of the satellites. Each of these interrelated trends is briefly reviewed below.

Growth in Launch Capability

As noted in Chapter 1, the most heavily used military space launchers have had their origins, not surprisingly, in the ballistic missile program. Long after such missiles as Thor, Atlas, and Titan ceased to be a significant part of the U.S. strategic inventory, they continued as the workhorses of the military space program. The thrust and, with it, payload capacity of these launch vehicles has been progressively increased by the use of more efficient engines and fuels, by strapping on additional boosters, and by use of special upper stages—the most famous being the Agena series. The Thor-Agena A that launched the early Corona satellites (1959–60) could put a payload of about 800 kilograms into low earth orbit (LEO).[26]

By 1965, the massive Titan IIIC and IIID boosters could put well over 13,000 kilograms in LEO or 1,500 kilograms into geostationary orbit. With the advent of the space shuttle, upwards of 29,500 kilograms can be placed in LEO. Moreover, such satellites launched by the shuttle can be designed to be retrieved or refueled.

Capabilities of the Payload

The growth in lift capacity has enabled the deployment of larger and multifunctional space platforms or, alternatively, the deployment of several satellites per launch. The first route is reflected in the reconnaissance/surveillance program. The initial Corona and SAMOS payloads were 880 kilograms and 1960 kilograms, respectively, with one camera system apiece (and probably some ELINT capability). The latest generation—the Big Bird and KH-11 satellites—weigh approximately 13,300 kilograms and have a multitude of sensors aboard (cameras, telescopes, infrared detectors, ELINT equipment, side-looking radar, and the like). Similarly, the DSP satellites use infrared telescopes for their early warning mission and also deploy nuclear explosion detectors.

The alternative practice of launching multiple satellite payloads has obvious economic benefits but is particularly useful for deploying space systems that operate in precise constellations. The U.S. Navy has been among the most consistent users of this technique for its geodetic, communications, radar calibration, ferret, and, more recently, ocean reconnaissance programs.

Frequency of Launches

Clearly, the most significant effect of the increase in launch capacity and overall payload capability has been to reduce the number of launchings required each year. Figure 2–2, showing the number of U.S. military satellites successfully launched into orbit from 1958 to 1981, clearly illustrates this trend.

Figure 2–2. U.S. Payloads Placed in Orbit, 1958–80.

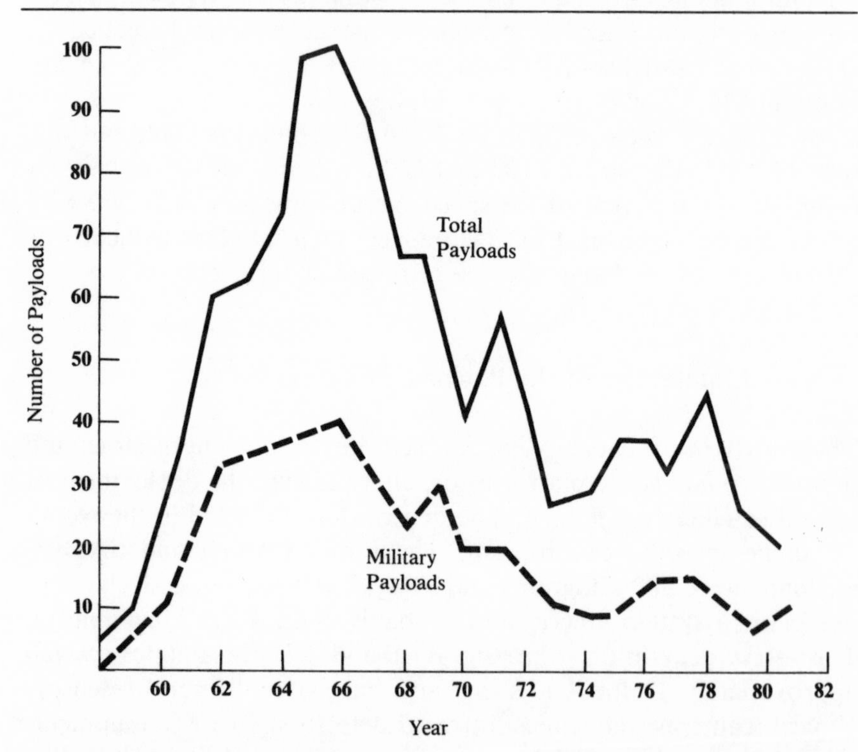

Sources: U.S. Congress, Senate Committee on Commerce, Science and Transportation, *Soviet Space Programs, 1976–80.* Washington, D.C.: Government Printing Office, 1982.

———. House Committee on Science and Technology, *Space Activities of the U.S., Soviet Union and Other Launching Countries and Organizations,* Congressional Research Service Report to the Subcommittee on Space Science and Applications, 98th Cong., 1st sess., Serial F. April 1983.

Table 2-1. Number and Weight of U.S. Payloads to Orbit, Five Year Intervals 1960–80.

Year	*Number of Payloads Orbited*	*Annual Total in Kilograms (est.)*[a]
1960	11	9,500
1965	46	38,600
1970	30	30,700
1975	13	36,000
1980	15	34,100

[a] Weights calculated from the roster of U.S. space launches in Appendix G of the source.
Source: U.S. Congress, House Committee on Science and Technology, *U.S. Civilian Space Programs, 1958–78, Vol 1.* Committee Print, Serial D. January 1981.

The decline in the number of orbited military payloads is often misinterpreted or unfavorably compared with the equivalent Soviet launch figures. Apart from the noted improvements to the capability of U.S. satellites, the decline in the U.S. launch rate can be put into further perspective by reference to the trend in the total annual weight of payloads launched into orbit. This is illustrated in Table 2–1, which compares these two indices at five-year intervals from 1960.

A further consequence of the increase in lift capability has been to extend the operational lifetimes of satellites, either through the use of higher orbits to avoid or limit the rate of orbital decay or by means of extra on-board propellants to sustain operations in low earth orbits. The development of more efficient and reliable power sources has also improved operational lifetime expectancies. Most communication, navigation, early warning, and some ELINT satellites had moved to higher orbits by the 1970s, and the four-to-five-day and two-to-three-week life expectancy of the original Corona and SAMOS satellites has risen to 250 days for the Big Bird and two years for the KH-11.

As the trend towards fewer, more capable spacecraft has developed, the level of U.S. military dependency on space systems has been increased still further. In addition to accurate strategic intelligence assessments and arms control verification, modern reconnaissance satellites provide the only real-time means for crisis monitoring; communications satellites are used to transmit over 70 percent of all peacetime long-haul messages; early warning satellites provide a potentially vital extra margin of time for crisis decision making;

weather satellites provide the only source for real-time global weather updates; and with the NAVSTAR GPS satellites, near perfect navigation and guidance information will be available to a multitude of military (and civil) users. These same satellites would provide the only real source for a post-attack assessment of the accuracy and damage caused by a nuclear attack.

The significance of this growing level of dependency can only be assessed by considering the effect that the temporary or permanent denial of satellite services, for whatever reason, would have on U.S. national security. The degree of dependency then becomes a function of the level of redundancy invested in space systems. This is measured by the availability of alternative space and earth-based systems or the ability to reconstitute the "lost" satellites. Without the provision of backup or replacement facilities, dependence on space systems becomes even higher. Although the United States can conceivably replace lost and damaged satellites or use alternative systems with relatively inconsequential *short-term* effects in peacetime, it is at least questionable whether the same could be achieved in a crisis or wartime situation.

CURRENT POLICY AND FUTURE ISSUES

The importance of this growing dependency on space systems has only become more pertinent with the increased interest by both the Soviet Union and the United States in the deployment of antisatellite weapons. This has raised the fundamental question as to whether it is prudent to encourage the development of a weapon that in itself threatens the very systems that are vital to U.S. national security and military operations generally. As illustrated above, U.S. policy in the past had been aimed at reducing the incentives for an ASAT competition in space because of the greater U.S. dependence on satellites. The Reagan administration's commitment to the development of a U.S. antisatellite system represents a significant departure from established policy. While ASAT development had been sanctioned by previous administrations, this had been defined either as a bargaining chip for negotiation with the Soviets or, if that failed, as a counter to the indirect threat posed by certain Soviet satellites. The rationale for a U.S. ASAT system was rarely linked with the object of negating the Soviet ASAT system. The Reagan administration, however, has not only shown little interest in reviving the stalled ASAT negotiations (although the arms control option has not been ruled out entirely) but, more significantly, has also argued that the deployment of a U.S. ASAT can *deter* the Soviet Union from inimical actions in space.[27] Although there may be some validity in the latter argument based on the Soviet Union's growing use of space for military purposes (discussed in chapter three), it is still questionable whether the superpowers' respective levels of dependency on space systems will ever be sufficiently balanced for deterrence to work. The potential benefits of an arms control regime that prohibits or places limits on dedicated antisatellite systems is still likely to be a more advantageous option for the United States and certainly a more attractive alternative than relying on a relationship that is likely to be escalatory, costly, and unstable during crises. Unfortunately, the momentum toward the deployment of antisatellite weapons by both the United States and the Soviet Union may already be irreversible.

In conclusion, the U.S. military exploitation of space has taken place in a virtual sanctuary without fear of direct interference. The benefits that have derived from this state of affairs have been considerable. If the exploitation of space enters a new phase that undermines this relatively benign operational environment, then clearly these benefits can no longer be guaranteed.

NOTES

1. It is fortunate that the end of U-2 flights over the Soviet Union had not occurred earlier as the first photographs from reconnaissance satellites were not recovered until August 1960 with Discoverers 13 and 14. It may not have been until 1961, however, that the quality of the satellite photographs were of the caliber of those from the U-2.
2. Although the satellite reconnaissance conducted by U.S. intelligence agencies does not in the strict sense constitute a military operation, for convenience here they are categorized as such.
3. The information in this chapter, unless otherwise stated, is drawn from the following: Paul B. Stares, *The Militarization of Space: U.S. Policy and the Development of Antisatellite Weapons* (Beckenham, Kent, U.K.: Croom Helm, 1984); Philip J. Klass, *Secret Sentries in Space* (New York: Random House, 1971); U.S. Congress, House, Committee on Science and Technology, Subcommittee on Space Science and Applications, *U.S. Civilian Space Programs, 1958–1978,*

Vol. 1. Committee Print, Serial D, January 1981; and Bhupendra Jasani, ed., *Outer Space: A New Dimension of the Arms Race*. (London: Taylor and Francis, Ltd., 1981).

4. This earlier policy goal has been stated in NSC Action 1553 of 21 November 1956 and in two subsequent letters from President Eisenhower to Soviet Premier Bulganin of January and February 1958.
5. NSC 5814/1, "Preliminary U.S. Policy on Outer Space," 18 August 1958; NASC, "U.S. Policy on Outer Space," 26 January 1960 (often referred to as NSC 5918).
6. For a detailed discussion see Gerald Steinberg, "The Legitimization of Satellite Reconnaissance: An Example of Informal Arms Control Negotiations," (PhD dissertation, Cornell University, New York, 1981.)
7. The White House, "Description of a Presidential Directive on National Space Policy," *Weekly Compilation of Presidential Documents* (20 June 1978): 1136–37; and "Fact Sheet, National Space Policy," *Weekly Compilation of Presidential Documents* (12 July 1982): 872–76. The reference to NSDD 42 can be found in U.S. Congress, House, Committee on Science and Technology, *Hearing Before the Subcommittee on Space Science and Applications on National Space Policy*. Committee Serial 143, 97th Cong., 2d sess., 4 August 1982, p. 13.
8. Laid down in NSC 5520, "U.S. Scientific Satellite Program," 27 May 1955.
9. John M. Logsdon, *The Decision to Go to the Moon: Project Apollo and the National Interest* (Cambridge, Mass.: MIT Press, 1970).
10. The NRO is apparently also subordinate to an interdepartmental panel known as the Executive Committee (EXCOM). See John Prados, *The Soviet Estimate: U.S. Intelligence Analysis and Russian Military Strength* (New York: The Dial Press, 1982), p. 109; and James Bamford, *The Puzzle Palace: A Report on NSA, America's Most Secret Agency* (Boston: Houghton Mifflin, 1982), p. 189.
11. Bamford, p. 190.
12. Logsdon, p. 73.
13. See Richard L. Garwin, "Bureaucratic and Other Problems in Planning and Managing Military R&D," in Franklin A. Long and Judith Reppy, eds., *The Genesis of New Weapons: Decision Making for Military R&D* (New York: Pergamon Press, 1980), p. 29.
14. Edgar Ulsamer, "Space Command: Setting the Course for the Future," *Air Force Magazine* (August 1982): 54–55.
15. Jeffrey T. Richelson, *United States Strategic Reconnaissance: Photographic/Imaging Satellites*. ACIS Working Paper No. 38, Center for

International and Strategic Affairs, UCLA, May 1983, p. 10. It is assumed here that successive generations photoreconnaissance satellites were given their KH designation on the basis of their chronological development.

16. SAMOS 2, launched in January 1961, did transmit some film although it is uncertain whether it was of much use.
17. The first successful launch of the second generation search and find satellites was on 18 May 1963. These had an average lifetime of twenty to twenty-five days. Further improvements to capability and operational duration were achieved by the third generation, starting 9 August 1966.
18. C. L. Peebles, "The Guardians: A History of the Big Bird Reconnaissance Satellites," *Spaceflight* 20:11 (November 1978); Desmond Ball, *A Suitable Piece of Real Estate: American Installations in Australia* (Sydney: Hale and Iremonger, 1980); Bamford (note 10), p. 201–2. Apparently there is no KH-10. See Clarence A. Robinson, Jr. "Soviets Accelerate Missile Defense Efforts," *Aviation Week and Space Technology* (16 January 1984): 15.
19. The first successful Air Force and Navy ELINT satellites were launched on 18 June and 13 December 1962, respectively. It is uncertain to what extent this service attribution was a cover for NSA operations.
20. The Navy would orbit upwards of six ferret satellites per launch.
21. See Ball, *A Suitable Piece of Real Estate.*
22. Farooq Hussain, *The Future of Arms Control, Part IV: The Impact of Weapons Test Restrictions,* Adelphi Paper No. 165 (London: International Institute for Strategic Studies, 1981), p. 62.
23. Richard Burt, "U.S. Plans New Way to Check Soviet Missile Tests," *New York Times,* 29 June 1979.
24. For example, Projects Score, Courier, and West Ford.
25. The NAVSTAR GPS program originated with the U.S. Navy's TIMATION experimental satellite project and the Air Force's Defense Navsat Development Program. The two service programs were merged in 1973.
26. The payload weights listed here are approximations taken from the Master Log of U.S. Space Flights, Appendix G of *U.S. Civilian Space Programs, 1958–1978* (note 3), pp. 1247–1332.
27. See U. S. Congress, Senate, Committee on Foreign Relations, *Hearings Before the Subcommittee on Arms Control, Oceans, International Operations and Environment on Arms Control and the Militarization of Space,* 97th Cong., 2d sess., 20 September 1982, p. 9. Testimony of then-Director of the U.S. Arms Control and Disarmament Agency Eugene Rostow.

REFERENCES

Ball, Desmond. *A Suitable Piece of Real Estate: American Installations in Australia*. Sydney: Hale and Iremonger, 1980.

Bamford, James. *The Puzzle Palace: A Report on NSA, America's Most Secret Agency*. Boston: Houghton Mifflin, 1982.

Burt, Richard. "U.S. Plans New Way to Check Soviet Missile Tests." *New York Times,* 29 June 1979.

Garwin, Richard L. "Bureaucratic and Other Problems in Planning and Managing Military R&D." In Franklin A. Long and Judith Reppy, eds. *The Genesis of New Weapons: Decision Making for Military R&D*. New York: Pergamon Press, 1980.

Hussain, Farooq. *The Future of Arms Control, Part IV: The Impact of Weapons Test Restrictions*. Adelphi Paper No. 165. London: International Institute for Strategic Studies, 1981.

Jasani, Bhupendra, ed. *Outer Space: A New Dimension of the Arms Race*. London: Taylor and Francis, Ltd., 1981.

Klass, Philip J. *Secret Sentries in Space*. New York: Random House, 1971.

Logsdon, John M. *The Decision to Go to the Moon: Project Apollo and the National Interest*. Cambridge, Mass.: The MIT Press, 1970.

Peebles, C. L. "The Guardians: A History of the Big Bird Reconnaissance Satellites." *Spaceflight* 20 (November 1978).

Prados, John. *The Soviet Estimate: U.S. Intelligence Analysis and Russian Military Strength*. New York: The Dial Press, 1982.

Richelson, Jeffrey T. *United States Strategic Reconnaissance: Photographic/Imaging Satellites*. ACIS Working Paper No. 38, Center for International and Strategic Affairs, University of California at Los Angeles, May 1983.

Robinson, Clarence A., Jr. "Soviets Accelerate Missile Defense Efforts." *Aviation Week and Space Technology* (16 January 1984): 14–16.

Stares, Paul B. *The Militarization of Space: U.S. Policy and the Development of Antisatellite Weapons*. Beckenham, Kent, U.K.: Croom Helm, 1984.

Steinberg, Gerald. "The Legitimization of Satellite Reconnaissance: An Example of Informal Arms Control Negotiations." Ph.D Dissertation, Cornell University, New York, 1981.

Ulsamer, Edgar. "Space Command: Setting the Course for the Future." *Air Force Magazine* (August 1982).

U.S. Congress. House. Committee on Science and Technology. *Hearing Before the Subcommittee on Space Science and Applications on National Space Policy*. Committee Serial No. 143. 97th Cong., 2d sess. 4 August 1982.

U.S. Congress. House. Committee on Science and Technology. Subcommittee on Space Science and Applications. *U.S. Civilian Space Programs, 1958–1978, Vol. 1.* Committee Print. January 1981.

U.S. Congress. Senate. Committee on Foreign Relations. *Hearings Before the Subcommittee on Arms Control, Oceans, International Operations and Environment on Arms Control and the Militarization of Space.* 97th Cong., 2d sess. 20 September 1982.

The White House. "Description of a Presidential Directive on National Space Policy." Press Release. *Weekly Compilation of Presidential Documents* (20 June 1978): 1136–37.

———. "Fact Sheet, National Space Policy." *Weekly Compilation of Presidential Documents* (12 July 1982): 872–76.

The White House. National Security Council. "Preliminary U.S. Policy on Outer Space." NSC 5814/1, 18 August 1958.

———. "U.S. Policy on Outer Space." NSC 5918, 26 January 1960.

———. "U.S. Scientific Satellite Program." NSC 5520, 27 May 1955.

3 SPACE AND SOVIET MILITARY PLANNING

Stephen M. Meyer

Much has been said and written about the enormous magnitude and scope of the Soviet military effort in space. Not only does the Soviet Union seem to be doing more in space than the United States, but it appears to be accelerating its activities, thus leaving the United States further behind. Indeed, the simple statistics so often cited seem to support this impression. In 1967 the United States boosted some 78 payloads into space in 57 launchings; the Soviets put up 74 payloads in 66 launchings. By 1980, however, the United States was placing fewer than 16 payloads in space (13 launches) in comparison to a Soviet effort of 132 payloads (89 launches). The trend continues basically unchanged: In 1982 the United States boosted fewer than two dozen payloads, while the Soviets launched some 119.

If the Soviet launch record for the past two decades is examined, it can be seen that during the 1970s more then 60 percent of all Soviet space payloads involved direct military missions (see Figure 3–1). More recently, the proportion of payloads dedicated to military missions has approached 75 percent. Another 10 percent provide both military and civilian services. A comparison of the distribution of military missions among Soviet payloads for three two-year periods –

Revised version of an article published in *Survival* (September–October 1983).

Figure 3-1. Soviet Payloads Placed in Orbit, 1960-82.

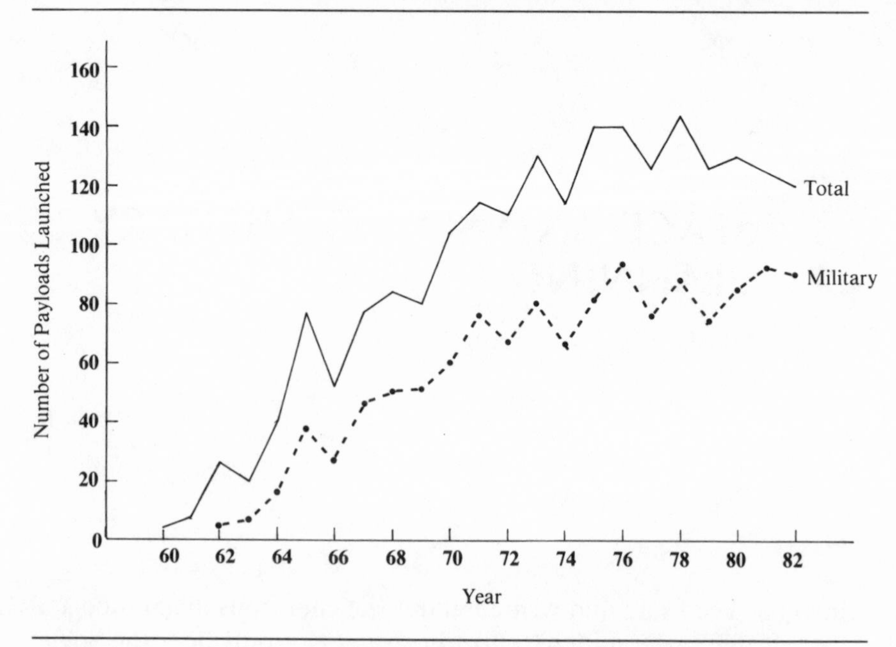

Source: Stephen M. Meyer, "Soviet Military Planning and the New High Ground," *Survival* (September-October 1983): 212. (Reprinted by permission of the publisher.)

Table 3-1. Distribution of Soviet Military Missions by Payload.

	1970-71 %	*1975-76* %	*1981-82* %
Military Photo-Reconnaissance	40	35	37
TACCOM	18	28	23
ELINT	7	6	4
Strategic Communication	6	9	8
Navigation	3	5	8
ASAT	4	2	2
ASAT Targets	3	2	1
Ocean Reconnaissance	2	4	7
FOBS	2	–	–
Early Warning	–	1	5
Radar Calibration	15	8	5

Source: Stephen M. Meyer, "Soviet Military Planning and the New High Ground," *Survival* (September-October 1983): 205. (Reprinted by permission of the publisher.)

1970–71, 1975–76, and 1981–82 – in Table 3–1 shows a furious amount of *launch activity* by U.S. standards. These are most specifically tied to intelligence collection (i.e., photoreconnaissance, electronic intelligence, ocean reconnaissance) and command and control of the armed forces (i.e., strategic communications, tactical communications, and early warning).

Though not included in the tabulation of military missions, major aspects of the Soviet manned space program are exclusively military. For example, it is well known that the Salyut space station has been outfitted and tasked in two distinct versions: one for military missions and one for civilian missions.[1] A wide range of reconnaissance, ELINT, and C^3 tasks and experiments have been performed by Soviet cosmonauts during their extended stays in space. During 1970–71 Soviet space platforms were manned for 33 days – 18 days for the Soyuz-9 and 24 days for the Salyut-1. In 1975–76 the Soviets logged 150 days – 93 days for Salyut-4, 49 days for Salyut-5, and 8 days for the Soyuz-22 capsule. In 1981–82 cosmonauts were in orbit for 286 days – 75 days aboard Salyut-6 and 211 days aboard Salyut-7 (the same crew remaining onboard for the entire 211 days).

While this descriptive information paints a fairly concise picture of Soviet space activities, it offers few insights into Soviet military capabilities in space or, more importantly, their value to Soviet military activities on earth. What military missions can the Soviets perform in (and from) space? How important are Soviet space capabilities to the performance of key military missions? How vulnerable are Soviet space capabilities to disruption? What can the Soviets do, unilaterally and through arms control, to enhance the capability and survivability of their space systems? This chapter directly addresses the first three of these questions.[2] My intent is to provide a foundation for discussion of the fourth.

SETTING THE CONTEXT

Most Western examinations of Soviet military programs – nuclear, conventional, naval, space – treat their subject outside the broader context of Soviet military doctrine, strategy, and planning. While isolating the "TNF balance," the "naval balance," or the "space balance" may simplify analyses, one runs a serious risk of exaggerating the relative importance of a given military force component. For example, navigation satellites may be critical to Soviet naval operations on the high seas, but the Soviet Navy may have many higher priority missions in wartime that preclude significant high seas operations. Moreover, even the highest priority navy missions may not figure prominently in the Soviet general war plan. Thus, while navigation satellites might be critical to a specific navy mission, they may have only minor significance in the broader context of Soviet military planning for theater war in Europe or the Far East.

Another important consideration is the fact that the primary "target" of Soviet politico-military power is Europe, not the United States. This is often difficult for American defense analysts to accept. Soviet military planners look upon the United States as the strategic rear area of Europe and, thus, the military demands/requirements for using Soviet forces (i.e., Soviet central nuclear forces) in the U.S. theater of military operations (TVD) are very different from those relevant to the use of Soviet forces (including nuclear forces) in the European theater. This is not meant to imply that the American TVD is less important than the European TVD to Soviet military planners. While the intrinsic values of these two theaters may be equal, Soviet politico-military goals—and hence approaches to military operations—in the two theaters can be, and are, quite different. In contrast to the desire to capture and preserve the economic and industrial infrastructure of Central Europe, Soviet military thinking about the deep strategic rear areas calls for the disruption and, if possible, the complete destruction of all military forces, economic and industrial potential, and political and administrative institutions in the United States.

A third consideration is the fact that the Soviet Union is a land power in both the European theater and the Far Eastern theater. This presents the Soviet military with many options and alternatives to space-based assets for carrying out those missions for which the United States must depend on space.

COMMAND AND CONTROL OF SOVIET MILITARY ACTIVITIES IN SPACE

Despite the best efforts of many Western specialists, the structure of the administration of Soviet space activities remains a riddle. There is some evidence, however, suggesting that a State Committee on Space Exploration oversees Soviet non-military activities in space, and coordinates long-term space planning between the Ministry of Defense and other state organs.[3] Regardless of the exact character of the State Committee on Space Exploration, the command and control of Soviet military activities in space is firmly in the hands of the Ministry of Defense, and administered by the various services of the armed forces. There is no Soviet space command.

The Strategic Rocket Forces

The Strategic Rocket Forces (Russian acronym: RVSN) administers most of the Soviet long-range (over 1,000 km) ballistic missile development programs and has operational control over all Soviet intercontinental ballistic missiles and intermediate range ballistic missiles (ICBMs and IRBMs). In fact, all current Soviet launch vehicles, save one, are derived from RVSN missiles. Given the RVSN's vast experience with developing and launching long-range missiles, it is not surprising that the RVSN conducts *all* Soviet space launches—military and civilian.[4] It may exercise administrative and/or operational control of the three Soviet launch facilities: Plesetsk, Kapustin Yar, and Tyuratam.[5] It may also control a number of ground stations and tracking ships that are part of their ballistic missile test support system.

The Air Forces

As far as is publicly known, the only significant role played by the Soviet Air Forces is in the area of cosmonaut training and space medicine. Of course, the many aeronautical science and engineering schools administered by the Air Force also contribute substantially to the Soviet military space program.

The Air Defense Forces

The Soviet Air Defense Forces (Russian acronym: VPVO) is believed to be divided into a number of branches, two of which have direct responsibility for space systems: the Troops of Antimissile Defense (VPRO) and the Troops of Antispace Defense (VPKO). Control of ground-based early warning radars is probably split between the

Radio-Technical Troops and the VPRO. Therefore, it is possible that either the VPRO subordinate command or the VPVO command itself has operational control over Soviet space-based early warning systems. (Alternatively, the General Staff may have direct control over those satellites). The VPRO may also control some aspects of Soviet research programs on space-based beam weapons for anti-missile defense.

Another branch of the VPVO, the VPKO, controls the Soviet ASAT program and most likely operates much of the Soviet ground-based space tracking network. It is also believed that the VPKO controls the Soviet ground- and space-based laser research program for ASAT applications.

The Navy

There are two military space programs that might fall under Navy control. The first and most obvious is the ocean reconnaissance satellite program. The second program involves the development and operation of navigation satellites. The Navy may also have operational control over the Soviet network of tracking and communications ships.

The KGB

The KGB is believed to have its own network of satellites that provide channels of communication parallel to those of the military command and control chain. Given that part of the mission of the KGB is to keep an eye on the military leadership, one can be sure that both the satellite systems and their ground links are run independently by the KGB.

It is also possible that the KGB controls some Soviet space-based reconnaissance systems, perhaps in cooperation with the GRU—the Soviet military intelligence organization.

SOVIET SPACE SUPPORT MISSIONS

Military Photoreconnaissance

Between 1975 and 1982 the Soviet Union launched an average of thirty-five military photoreconnaissance satellites (MPRSATs) annually.

(In contrast the United States launched two to four a year during the same period.) This amount of activity seems to imply both tremendous Soviet interest in and dependence on MPR satellites. In fact, the data are quite misleading. Until recently, Soviet MPRSATs exhibited maximum orbital (mission) lifetimes of about ten to thirteen days, after which time they would reenter the atmosphere and return their exposed film cargoes to the Soviet Union. With such short mission times, a minimum of thirty satellites must be launched each year merely to maintain a single operational MPRSAT in orbit on a daily basis—even assuming that there were no major satellite malfunctions.

In recent years, the Soviet Union has begun to experiment with longer duration satellites and film-return capsules. Nine of the thirty-five MPR payloads launched in 1982 had mission longevities over thirty days—the longest was fifty days. Yet, even with these longer duration satellites, only 680 MPR mission-days were logged during 1982, meaning that the Soviets averaged no more than two MPRSATs available per day over 315 days, and only one operating per day over the remaining fifty days.[6]

Thus, it is obvious that the high annual launch rate of Soviet MPRSATs is a poor (and misleading) indicator of the Soviet "presence" in space or Soviet MPR capabilities from space because the Soviet Union has purposefully adopted a high launch rate/low mission longevity engineering approach to the MPR mission.

Apparently, the Soviet Union has chosen to fly separate satellites for high-resolution (close-look) versus medium- and low-resolution (area) photoreconnaissance missions. The high-resolution satellites are maneuverable and tend to have orbits with apogees around 350 kilometers and perigees as low as 165 kilometers. The medium- and low-resolution satellites travel in orbits that are more circular, with perigees above 225 kilometers. The use of two different satellite systems for close-look and area reconnaissance increases further the launch rate demand for the MPR mission.

The majority of MPRSATs appear to fly high-resolution missions. Beyond general reconnaissance of the United States and Western Europe, studies of ground tracks reveal a particular Soviet interest in following Third World military conflicts, for example, engagements in the Middle East, the Persian Gulf area, and South Asia.[7] In the past, Soviet use of film-return *satellites* (rather than capsules) has meant that a replacement MPRSAT had to be launched every time that "current" photoreconnaissance data were required, and so there is a fairly high correlation between Third World crises and extraordinary launches of Soviet MPRSATs. The recent introduction of high-resolution systems employing film-return capsules will weaken this correlation.

While space-based MPR systems are valuable during peacetime and in crises for monitoring current events and developing trends (particularly outside the European theater), reconnaissance planning for the Soviet primary contingency—war in Central Europe and/or the Far East—appears to be premised on the use of theater-based air, ground, and naval reconnaissance systems.[8] To be sure, the Soviet strategic command would find MPR data from space useful for monitoring the strategic disposition of forces and supplies in the rear areas of NATO Europe. However, normal weather patterns over the region—cloud cover in particular—call into question the relative utility of space-based MPR systems over air and ground MPR systems.

Even in the context of a U.S.-Soviet intercontinental nuclear war, one finds no evidence of a Soviet "shoot-look-shoot" strategy requiring post-strike MPR capabilities. Instead one finds a clear tendency towards a multisalvo artillery fire concept that does not require near real-time target damage assessment.[9] That is to say, the Soviet strategy for nuclear strikes against the United States (which, again, it views as the strategic rear of Europe) does not require space-based reconnaissance. Needs for strategic intelligence during crises and confrontations may be met by Soviet GRU and KGB operatives stationed in the United States and Europe for just such purposes.

In instances of "limited" Soviet military engagements around its periphery, air-, ground-, and sea-based reconnaissance systems are available to provide real-time information for battle management and troop control—something that Soviet MPRSATs do not provide. From the perspective of the operational commanders, these reconnaissance systems are preferable to space-based systems because they are under direct control, they are responsive to theater requirements and demands, and they provide real-time information.

While Soviet MPRSATs are all in low orbits (under 400 km) and hence will be vulnerable to attack by the planned U.S. F-15/ASAT, the Soviet space-based MPR capability is partially protected by its high launch rate capacity. That is to say, the fairly continuous production of both satellites and boosters in peacetime (the A-2 booster is not taken from stocks of deactivated ICBMs) suggests a Soviet potential to replace disabled or malfunctioning MPRSATs within a few days, and with little concern for spares. Of course, the vulnerability of the two main Soviet launch complexes, Plesetsk and Tyuratam, raises doubts about the efficacy of reconstituting the space-based MPR mission during major war. These would be easy targets for U.S. ICBMs, SLBMs, and bomber-launched cruise missiles, as well as British and French nuclear forces.

Electronic Intelligence

The Soviets maintain a six-satellite constellation to support the global ELINT mission. These passively collect data on the content and characteristics of transmissions and emissions from the many sources of electronic signals that emanate from modern armies, including radar systems, radio communications systems, data relay systems, C^3 facilities, and forces in the field. Over the last several years the Soviet Union has boosted three or four ELINT payloads annually, which, given their exhibited mean mission longevity of about 600 days, is just what is required to keep the basic constellation fully operational.[10]

The importance of Soviet space-based ELINT missions is similar to that of its MPR missions. Given the high density electronic environment of any major conflict and especially the need to coordinate electronic support measures (ESM) and electronic countermeasures (ECM) in real time, the primary ELINT missions are to be handled by air, ground, and naval systems.[11] In this respect, the main contribution of space-based ELINT gathering occurs during peacetime—when data on opposing forces are collected and analyzed, and countermeasures are developed. Thus, the loss of the ELINT constellation is not likely to affect Soviet military operations during war in a major way.

Travelling in near circular orbits at 600 km, these satellites would be vulnerable to attack by the planned U.S. F-15/ASAT.

Command, Control, and Communications

The Soviet Union maintains three distinct satellite constellations for military C^3. These service a variety of needs from tactical/theater communications to strategic command and control to strategic intelligence relays.

The first two constellations are deployed at altitudes of 800 kilometers and 1,300 kilometers. The lower constellation is comprised of three satellites in planes spaced 120 degrees apart. The second constellation is made up of twenty to thirty smaller satellites all moving in the same basic plane. Both constellations have identical inclinations.

These two groups of satellites are usually identified with tactical or theater communications missions since they do not appear to have the range to satisfy strategic communications requirements. While they do indeed serve tactical needs, they also play an important strategic role.

These satellites employ "store/dump" relay designs, meaning they receive information from an earth-based transmission point, store the transmission onboard, and then retransmit the message later on command from the appropriate receiving stations.[12] In this way, strategic ballistic missile submarines (SSBNs), dispersed command posts, and KGB/GRU agents operating around the world can use fairly low-powered transmitters to send messages and data back to Moscow. These communications can have strategic significance. However, since these satellites orbit at relatively low altitudes, transmissions to stations outside the immediate theater will experience relay delays from tens of minutes to hours—not true real time C^3.

In recent years, the Soviet Union has launched between twenty to thirty of these low-altitude communications satellites annually. Here again, the initial impression is one of great interest in, if not dependence on, space. However, the explanation for this high communications satellite launch rate is based more on geography, geometry, and engineering than on a special appreciation for the new high ground. In particular, the high latitudes of the Soviet Union and its military control centers, and the geographical disposition of its armed forces, creates a need for well-populated constellations of communications satellites in order to maintain a (fairly) reliable and (usually) continuous worldwide space-based C^3 network. There is also the problem of the message storage capacity of the satellites, as well as individual systems reliability.

Data on orbital replacement suggest a mean mission lifetime of 500 days for a given TACCOM satellite. Assuming that the Soviets wish to avoid a catastrophic failure of the TACCOM system, and given that the satellites of the more populous constellation are launched in octuples, maintaining the TACCOM network requires an average of four launches a year. This would involve two single launches for the first constellation and two octuple launches for the second constellation. Consequently, the Soviets must launch at least eighteen

TACCOM satellites per year merely to maintain a basic worldwide tactical C^3 capability. Naturally, unexpected failures would drive up the launch rate.

Space support for continuous and real-time strategic C^3 is provided by a third constellation: the Molniya satellites. The Molniya-1 constellation—which probably carries the bulk of the military traffic—consists of eight satellites each in its own orbital plane, spaced 45 degrees apart. (The Molniya-3 satellites are used for civilian communications, having replaced the Molniya-2s.) The Molniya orbits are highly eccentric (500 km by 40,000 km) so as to permit reliable and continuous coverage over the Soviet Union. Given the location of Soviet launch sites, these highly elliptical orbits also permit much heavier payloads to be placed into orbit, compared to what could be placed into equatorial stationary orbits. This may be an important consideration if the limits of Soviet space technology do not permit the level of microminiaturization achieved in the West.

Over the last several years Molniya-1 mission lifetimes have averaged about 900 days, suggesting that at least three replacement launches are required annually to maintain the basic constellation. In fact, we observe three or four replacement launches per year. Here again geography, geometry, orbital mechanics, and Soviet engineering conspire to demand a high launch rate.

How essential are Soviet space-based communications systems to Soviet military capabilities? In peacetime, they are a convenient if not important element in the command and control of Soviet forces deployed in Eastern Europe, the Far East, and on the high seas. But perhaps more importantly, space-based communications systems may hold special significance for Soviet military involvement in the Third World.

Historically, the very strict and centralized command and control structure of the Soviet armed forces, which stretches up to the political leadership, imposed great caution and inertia on military activities as distances from Moscow increased. However, the advent of C^3 satellite networks and mobile ground terminals raises the possibility of extending tight and centralized control of Soviet military activities well beyond traditional areas of Soviet military activity. For example, there are reports that recent Soviet SA-5 air defense missile battery deployments (involving Soviet operators) in Syria are accompanied by a satellite C^3 link to Moscow. In this respect, space-based communications systems may serve to increase the prospects for Soviet military involvement in the Third World. Undoubtedly, these satellite constellations provide valuable support to KGB and GRU operations in the Third World as well.

If there were any single wartime C^3 link that might benefit from the existence of, and hence be sensitive to the loss of, satellite communications it would be the strategic communications link between the General Staff and the Front (or strategic direction/theater) commanders. However, there does not appear to be much in the way of evidence to suggest that space-based systems have assumed this role. While ground terminals may exist at major Ground Forces command posts, the primary routes remain redundant ground-line systems and ground- and air-based radio relays, which continue to receive priority attention. The extensiveness of Soviet earth-based C^3 systems suggests that the space-based C^3 systems are backups.

Nor is there reason to believe that satellite communications will play an important role in intratheater battle management. First, Soviet C^3 principles have always been oriented in a top-down fashion, which discourages communication from lower to higher formations or between troop formations at the same level. (In the latter instance, communications are routed through higher formation communications posts.) Second, the continued poor state of basic radio communications management at the unit level in the Ground Forces does not speak well for the prospects of satellite networking. In fact, nonradio means of communication continue to play a central role in Soviet unit level C^3.

One might imagine that the Soviet Navy would suffer from the disruption of space-based communications. However, in time of great crisis and confrontation the evidence suggests that Soviet naval forces will be pulled in near to the land theaters for purposes of homeland defense, closing off Soviet SSBN sanctuaries and interdicting enemy sea lines of communication (SLOC) adjacent to the theater. Here a number of ground and air radio communications relay options are available. Soviet SSBNs will not have to worry about C^3 links to the strategic command in Moscow. Soviet naval aircraft and surface forces deployed around the SSBNs are exercised to serve as relay points.

Though the two lower altitude communications satellite constellations probably will be within the range of the U.S. F-15/ASAT, they are fairly resistant to ASAT interceptor attacks owing to the large number of satellites in the two constellations. While the destruction of a dozen TACCOM satellites would most likely create communications holes for periods of time, the store-dump design of these satellites suggests that basic communications links would remain intact. Moreover, if Soviet launch facilities are not attacked, much of the damaged constellation could be reconstituted within several days by standard octuple satellite launches. Active jamming of Soviet satellite relays and transmissions within the theater probably represents a much greater threat to these space-based communications capabilities.

The destruction of the Molniya constellation would be more complicated, owing to the highly eccentric orbits and the placement of their ascending nodes. These orbital characteristics place them beyond the range of planned U.S. ASAT interceptors (given current basing and deployment schemes) during passage over the northern hemisphere. In any case, attacks against a half-dozen or so Molniya satellites would still leave enough of the system intact for large amounts of (well-timed) communications traffic.

Early Warning Systems

It has taken the Soviet Union more than a decade to complete the dedicated constellation for its space-based ballistic missile early warning (EW) system, a project plagued by a continuing series of satellite failures. Soviet EW satellites are deployed in a constellation of nine planes, spaced at 40 degree intervals, with one satellite to a plane. The orbits are highly eccentric, 500 to 40,000 kilometers, and synchronized with the earth to provide continuous coverage of the United States. From the observation points near apogee simultaneous coverage of missile fields and direct transmission to the Soviet Union is possible. Within the last two years the ascending nodes of the EW constellation apparently have been shifted 30 degrees east, resulting in an adjustment to the limb views over Chinese and U.S. ICBM fields.[13]

The average mission lifetime for Soviet EW satellites launched over the last several years appears to be about 600 days, requiring four or five launches a year in order to maintain the basic constellation. Observed launch rates have been four EW satellites per year.

Since it has taken the Soviet Union so long to complete its EW constellation it seems safe to conclude that it has other means of early warning with which it feels comfortable. Rudimentary EW sensors are believed to be aboard other Soviet satellites in high orbit (e.g., Molniyas). Ground-based radars (both line-of-sight and over-the-horizon designs) also are deployed. Soviet KGB and GRU agents, monitoring activities near nuclear weapons bases, are expected to provide some strategic warning. Nevertheless, the EW satellites probably do provide the longest warning time of ballistic missile launches from the United States, as well as some systematic coverage of likely SLBM launch areas.

Since their orbital characteristics are similar to those of the Molniya satellites, destroying the Soviet EW satellites would be very difficult given current deployment schemes for the U.S. F-15/ASAT. Spoofing or blinding by laser might be more effective countermeasures.

Ocean Reconnaissance Missions

The ocean reconnaissance mission is one of several examples where the USSR's military use of space is dictated by the logic of its perceived military requirements—in contrast to the often voiced notion that the Soviets merely follow the U.S. lead in the military use of space. Soviet ocean reconnaissance satellites are designed to provide location, tracking, and other intelligence data on foreign naval forces.

Two basic satellites are used: active radar-equipped ocean reconnaissance satellites (RORSATs) and passive sensor-equipped ELINT ocean reconnaissance satellites (EORSATs). RORSATs use small nuclear reactors to power the radars that locate and track ships at sea. When working properly, RORSATs appear to have mission lifetimes between 90 and 120 days operating at altitudes of about 260 kilometers, after which time their nuclear power source is boosted to high orbit (to prevent reentry). These satellites received some notoriety in the past several years when two malfunctioned and reentered the atmosphere, one crashing in Canada and creating local radioactive contamination.

EORSATs are less well known but in fact spend much more time working in space than RORSATs. EORSATs carry solar panels to power their sensors. They orbit at about 440 kilometers and seem to be designed for longer missions of at least 180 days duration.[14] While both satellite systems have suffered many operational problems, it appears that the Soviets are attempting to keep one RORSAT and one EORSAT in orbit at all times.

In peacetime, ocean reconnaissance satellites are undoubtedly valuable to the Soviet military. They provide much useful intelligence on foreign naval force deployments, naval tactics and operations, offensive and defensive capabilities and weaknesses, and so forth. Likewise, they could prove very useful in superpower crises (similar to the Cuban Missile Crisis) or third party and "limited" conflicts. Thus, the Soviet investments are not at all unreasonable. However, there is reason to doubt that the destruction of ocean reconnaissance satellites would have much of an impact on Soviet military capabilities or the main thrust of Soviet operations during a major U.S.-Soviet war.

To be sure, ocean reconnaissance satellites could be useful for locating convoys on the open ocean or aircraft carrier battlegroups posing threats to Soviet forces and territory.[15] Yet, it is clear from Soviet naval writings and exercises that deep ocean interdiction of the SLOC is a lower priority mission of the Soviet Navy and that, despite Admiral Gorshkov, naval missions (aside from SSBN operations) have the lowest priority in Soviet military planning.

In the context of any likely U.S.-Soviet military confrontation in which the destruction of space assets might be contemplated, most if not all Soviet naval forces will be operating in the ocean theaters adjacent to Europe and the Soviet Union. To the extent that SLOC interdiction is practiced, it will be the mission of Soviet Naval Aviation and groups of attack submarines, supported by long-range ocean reconnaissance aircraft providing data on convoy routes. It is likely that most attacks would occur closer to the continent where other Soviet naval reconnaissance aircraft can operate in conjunction with Backfire bomber raiding parties. Then too, more consistent with Soviet military strategy is the notion of destroying enemy SLOC by destroying the loading and/or unloading endpoints, rather than severing the connecting route. While data from ocean reconnaissance satellites might provide useful supplementary information, they would not be necessary for the success of Soviet naval missions.

Similarly, Soviet operations against carrier battlegroups would take place in areas where satellite reconnaissance would be useful but not essential. Air, surface, and subsurface reconnaissance means will be the main sources of data on the location and movement of carrier groups. A good illustration involves the Falkland Islands War. At the time that the British task force set sail for the South Atlantic (5 April 1982), no Soviet ocean reconnaissance satellites were in orbit. Only on 29 April did the Soviets launch an EORSAT. A RORSAT was finally orbited on 14 May—almost one-and-a-half months later. The reconnaissance burden was borne by fleet and air elements of the Soviet Navy.

Disruption of Soviet ocean reconnaissance satellites would not be overly difficult. RORSATs orbit at fairly low altitudes and their radars act as beacons. A U.S. ASAT system with a radar homer would have little trouble locating its prey. Though Soviet EORSATs orbit at somewhat higher altitudes and do not emit intense signals, they would be quite vulnerable to ASAT attack using infrared seeking. It also should be noted that electronic countermeasures can be employed to jam, deceive, and otherwise confuse Soviet ocean reconnaissance satellites. Indeed, spoofing may prove to be more efficient and effective than destruction of ocean reconnaissance satellites, especially if the data result in the misdirecting of other Soviet ocean reconnaissance means. Then too, spoofing would be less risky and more readily applicable in a crisis or limited conflict situation.

Navigation Satellites

The Soviet Union employs two constellations of navigation satellites (NAVSATs), comparable to the U.S. Transit system (see Chapter 5), that are maintained at altitudes of about 1,000 kilometers. The first constellation is comprised of six satellites, each in a separate plane spaced 30 degrees apart. The second contains four satellites spaced in separate planes 45 degrees apart. As mission lifetimes are of the order of 680 days for these satellites, the observed launch rate of six or so per year is about the minimum necessary to keep the constellations functioning.

Recently, the Soviets have begun orbital testing of a new generation of NAVSATs, dubbed GLONASS, akin to the U.S. NAVSTAR system (also discussed in Chapter 5).

Soviet dependence on NAVSATs, as with other space support systems, is a matter of context. In peacetime, NAVSATs greatly simplify long range military missions in distant theaters. However, there is reason to argue that, in a shooting war, Soviet military capabilities would not suffer for the lack of space-based navigation. In the context of the air-land battle in the theater, the Soviet Air Force has provided for a wide range of ground control points on the battlefield and in the rear areas, and radio locator beacons are also deployed.

The one service that would benefit most from NAVSATs is the Soviet Navy. But, as has been noted several times, priority missions for the surface fleet are all close in to the Soviet homeland, and the SSBNs will be operating from protected waters; they will know where they are. Moreover, there are a number of ways to provide supplementary navigation data, including preplaced ocean bottom navigation points that could serve as locational references for updating SSBN guidance computers.

Current Soviet NAVSATs orbit at altitudes that may be within the operational limit of the U.S. F-15/ASAT, but the new Soviet NAVSAT system will orbit at about 18,000 kilometers—well beyond the reach of that system.

SPACE WEAPONS PROGRAMS

With all the attention that the Soviet ASAT program has attracted, it is often forgotten that there were earlier Soviet efforts to develop space weaponry. It might be useful to review briefly these early programs to provide a context in which to examine the more contemporary Soviet efforts.

Nuclear Weapons in Space

There is evidence to suggest that, during the 1960s, the Soviet military began development work on orbital bombardment systems (OBS) and fractional orbital bombardment systems (FOBS).[16] The former was a nuclear weapon designed for placement in low earth orbit where it would remain until commanded to reenter the atmosphere and strike a target. The latter also was a nuclear weapon designed to follow an orbital trajectory between silos in the Soviet Union and targets in the United States, but it would reenter the atmosphere and attack its target before completing a full revolution. Coming in at low altitudes (compared to ballistic trajectories) OBS and FOBS could sneak in under line-of-sight radars and also attack from directions other than over the North Pole, where the U.S. early warning radars were concentrated.

Apparently, the OBS project was linked to the ill-fated SS-10 ICBM program. It is unclear whether any OBS tests were ever conducted, but the SS-10 and OBS development programs were canceled in the mid-1960s. OBS were banned by the Outer Space Treaty of 1967 (discussed in Chapter 7).

The FOBS program was tied to the SS-9 ICBM (F-1-r launch vehicle). Of the eighteen identified FOBS tests between 1966 and 1971, fifteen appear to have been successful. Although a Soviet FOBS is believed to have been operational between the late 1960s and early 1970s, the deactivation of the SS-9 ICBMs in the late 1970s and the lack of subsequent tests with new missile bodies implies that a FOBS is not operational today.[17] FOBS was banned in the SALT II Treaty.

ASAT Interceptor Program

Of all the Soviet military space projects, none has generated as much concern in the West as the ASAT interceptor program. The data on Soviet ASAT tests suggest that there have been two program test series: 1968–71 and 1976–present (see Figure 3–2). The first test series involved a radar homing satellite that used co-orbital intercept to approach a designated target satellite. The test record suggests a fairly successful series: F S F S S S S. It is possible that this system attained limited operational status (using the F-1-m launcher: a derivative of the SS-9). This initial period of ASAT testing overlapped the development period of the Soviet FOBS.

Some five years passed before further ASAT testing took place. The second series of tests incorporated two notable changes. First, a "pop-up" intercept technique was tried, offering the prospect of quick approach to and destruction of the target. But performance of the radar homing system in this mode was less than inspiring: F S F F F S S. Second, a new advanced homing system was tested, believed to employ optical tracking. However, the test data reveal that these innovations resulted in complete performance failure: F F F F F F.[18]

Certainly a critical aspect of any ASAT system is the capability to carry out a rapid sequence of launches involving numerous interceptor satellites, which culminates in the destruction of many independent targets within a short period of time. This has yet to be attempted in any Soviet ASAT test. Moreover, Soviet ASAT tests

Figure 3–2. Soviet R&D Test Patterns, 1964–82.

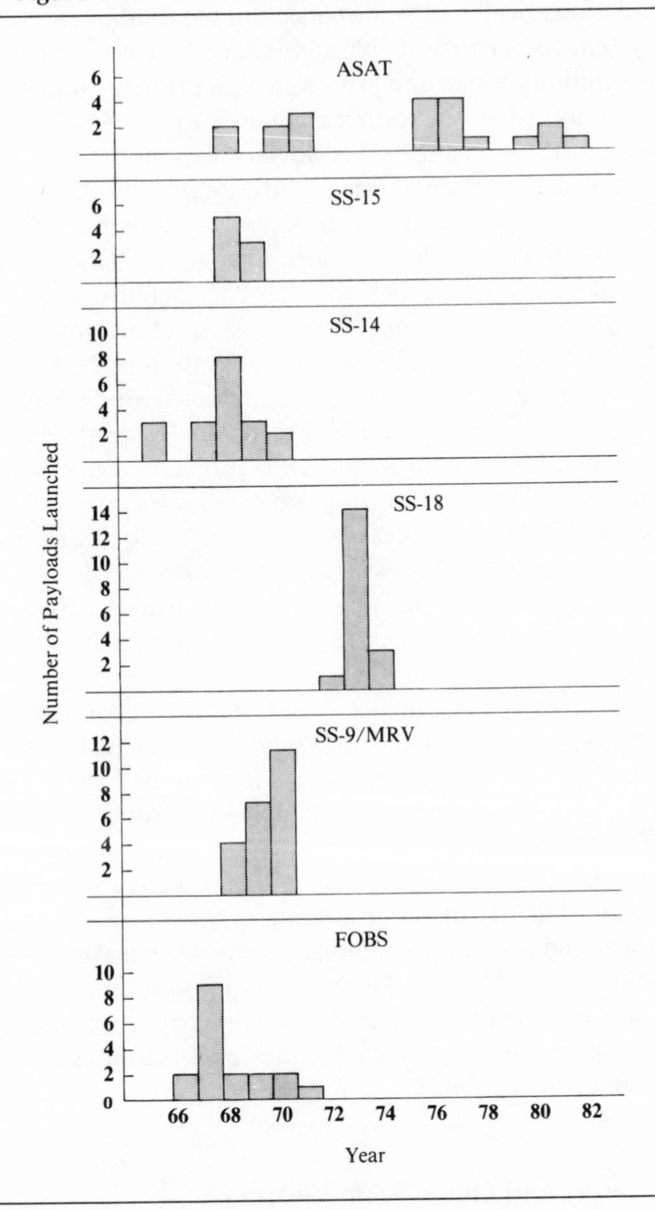

Source: Stephen M. Meyer, "Soviet Military Planning and the New High Ground," *Survival* (September–October 1983): 212. (Reprinted by permission of the publisher.)

have maintained a fairly narrow intercept regime in terms of altitudes and inclinations, which calls into question the fundamental capability of the system to perform its basic mission. Nor does it suggest that the Soviet military looks upon this as a high priority military mission.

Interestingly, when the testing frequency of the ASAT program is compared to the patterns of other Soviet weapons programs a noticeable difference is apparent (see Figure 3–2). The ASAT program seems to have been pursued at a leisurely pace, one that is not characteristic of the Soviet military's priority programs. One hypothesis is that, during the 1968–71 period, the Air Defense Forces' ASAT project was unable to compete for booster allocation against the Strategic Rocket Forces' SS-9 ICBM program and FOBS program, all which use the same basic "F-type" launcher. Within the Air Defense Forces, the program may have had to compete for system development funds against the concurrent ABM project.[19] During the second period, the ASAT interceptor program may have competed for funds with the Air Defense Forces' laser and particle beam programs, and/or the new investment efforts in anticruise missile systems, *and lost*. In any case, it does not appear that the Soviet ASAT program has even been accorded the priority commensurate with the attention it has received in the West.

How susceptible is the Soviet ASAT system to countermeasures? ASAT interceptors that use the co-orbital approach technique are vulnerable to direct ascent interception (that is, to the U.S. F-15/ASAT) because of the time they spend catching their target. The switch to "pop-up" interception by the Soviets, however, casts serious doubt on the prospects for destroying Soviet ASAT interceptors by conventional means. The use of radar homing in Soviet ASAT interceptors presents a number of opportunities for countermeasures, including target satellites designed to change orbits upon sensing radar illumination. Yet, here again, the possible move to optical tracking by the Soviets would greatly complicate the task of developing satellites that could "defend themselves."

Lasers and Other Beam Weapons

Little information about Soviet beam weapon programs for space applications is available publicly. Allegedly the Soviet Union is pursuing the technology of ground-based lasers for use against space objects, and the technology of space-based lasers (and perhaps particle beams) for use against space objects and aerial targets. To be sure, the Soviet military has long been aware of the potential application of beam weaponry for space-based weapons.[20]

Ground-based laser systems have the advantages of not being constrained in size, being readily accessible for servicing and repair, and being deployed on home territory, safe from adversary attack except during a general war. Their biggest problems, however, are that they have limited attack windows against enemy satellites—prospective targets must pass overhead. Moreover, ground-based lasers are "fair weather" weapons—unlike the U.S. Postal Service they will not work in rain, sleet, snow, or cloud cover.

Space-based systems have a much greater field of view, and their beams would not be subject to severe atmospheric attenuation. However, the technological problems associated with space-based laser systems (related to beam generation, tracking, pointing, platform dimensions, and the like) are very severe (see Chapter 6).

Given Soviet military research and development style for weapons that are new in principle, it would not be surprising if the Soviet Union were to deploy beam weaponry on the ground and in space before the United States. Historical examples of Soviet "firsts" include the first deployed ICBM, the first deployed SLBM, and the first deployed ABM. In each case, the Soviet Union deployed what the United States would have considered an immature technology, which had little or no capability to accomplish its task. The Soviet goal, however, was to get the basic weapon into the field and then work on developing its performance capabilities. Thus, it will not be surprising if the forecast Soviet directed energy weapons enter service with no military utility. Standard operating procedures within the military industries suggest that at least a decade of in situ modification and innovation will be required before a functional weapon finally emerges.[21]

While it is difficult to assess from open source information the state of Soviet technology pertinent to a ground-based weapon, some observations can be made about the prospects for a Soviet space-based laser system in the near future. The operational problems of Soviet space platforms are well documented. The truncated lifetimes of RORSATs suggest power failures, electronic failures, or attitude control limitations. Problems onboard EORSATS and EW satellites suggest more of the same. The most automated of all Soviet space hardware—lunar and interplanetary probes—have had the most dismal success rate of all. Indeed, if one examines the success/failure rate of Soviet space probes one finds that the lunar probes have the best record, the Venus probes have a poorer record, and the Mars probes have the worst record of all. In essence, the Soviets have exceptional trouble keeping their automated space probes functioning for more than a few months. The combination of complexity, longevity, and reliability seems to have been beyond Soviet technical capabilities. The technical demands of a laser ASAT system will be far greater.

Interestingly, similar failure modes have been observed on the Salyut space stations. A major aspect of cosmonaut station keeping activities (and Soyuz and Progress replenishment runs) has involved the repair and replacement of expended components.[22] Indeed, the unmanned performance of both Soyuz spacecraft and the Salyuts clearly demonstrates that cosmonauts are not needed for normal operation but for maintenance. Thus, if the Soviets are serious about placing a laser weapons system in orbit, they will most certainly have to deploy it as a module of a manned platform.

CONCLUDING OBSERVATIONS

The observed high activity of the Soviet military space program, as gauged by launch rates and payload emplacement in orbit, is not the product of some special Soviet insight or interest in the new high ground. Rather, it reflects the combined effects of geography, geometry, technology, and engineering style that make that level of activity necessary to maintain a minimal Soviet military mission presence in space. As the previous discussion shows, the Soviet Union needs to boost at least seventy satellites annually in order to maintain the basic capabilities in space that the United States enjoys with far fewer launches. Making allowances for replacements for unexpected failures, this leaves a residual of about a dozen payloads for manned space efforts, test programs, and systems development. The Soviets may indeed outspend the United States in space two to one, but they probably get little—if any—extra in return.

While the high launch rate required by the Soviet Union may be wasteful in peacetime, it suggests a resiliency that could prove useful should satellites become the targets of military attack during crises or war. On the one hand, the large number of satellites used in Soviet constellations and the high launch rates used to maintain them make Soviet military space capabilities fairly insensitive to discrete kills by limited ASAT attacks. Clearly, the relative vulnerability of Soviet space-based military systems will depend on the *quantitative* level of the ASAT threat posed by the United States. On the other hand, high Soviet peacetime launch rates also suggest a surge capability to reconstitute satellite populations following enemy attacks. For the United States, active disruption of Soviet space systems by jamming, spoofing, and other forms of countermeasures may be more efficient, effective, and timely than direct ASAT attacks on Soviet spacecraft.

For the most part, Soviet military investments in space appear to be oriented for peacetime and crisis management requirements—not wartime battle management. While the disruption or loss of Soviet space systems would severely affect Soviet military activities in the Third World, this is not the kind of confrontation where the United States or the Soviet Union is likely to consider attacking each other's space assets. In contrast, the loss of space assets might create inconveniences to Soviet military operations in the land theaters adjacent to Soviet territory, but such a loss would not be militarily significant. Indeed, from this perspective, *Soviet satellites are merely auxiliary systems—not primary systems—for performing military support missions.*

Should a shooting war erupt in space, Soviet military planners obviously would prefer an outcome in which Soviet space assets remained intact and U.S. satellites were destroyed. The Soviet Union does value its military space capabilities. Barring this, their next preference is likely to be a situation in which neither side has satellites in orbit because the asymmetries would favor the Soviet Union, at least over the short term. In other words, the Soviet military may place higher value on the destruction of U.S. space systems than on the preservation of its own. The next best situation would see both sides with space assets intact. The least desirable situation, from the Soviet military's perspective, would have U.S. satellites functioning and Soviet satellites destroyed.

This suggests that the United States will not be able to deter a Soviet ASAT attack on U.S. satellites by posing an analogous threat to Soviet space assets if the two sides have roughly equivalent ASAT capabilities. If, however, U.S. abilities to disrupt Soviet space operations are perceived by the Soviet military leadership to be significantly better than corresponding Soviet capabilities, then the Soviet Union would have to fear that any hostile action against U.S. space systems might rapidly lead to the worst possible situation: no substantial degradation of U.S. space assets but the annihilation of its own satellites. Thus, even though the Soviet Union may be less dependent than the United States on space-based military support systems, perceived asymmetries in ASAT capabilities favoring the United States could deter hostile Soviet actions against American satellites. Then too, if U.S. ASAT technologies suggest that the best the Soviets could hope for is their third preference, their interest in ASAT arms control is likely to grow. (This is not a bargaining chip gambit, just simple military logic).

There are two Soviet military space efforts that seem to be acting as stimuli for increased U.S. investment in space weapons missions: Soviet ASAT efforts and suspected Soviet beam weapons programs. With respect to the former, it was noted that the Soviet ASAT interceptor program appears to be a low priority project. After fifteen years of testing different versions of the interceptor system there are no indications that the Soviet Union is trying to create any militarily significant antisatellite capability through this approach.

With respect to beam weapons in space, Soviet military research and development style leads one to expect that the Soviet Union could well be the first to orbit a laser weapon-bearing satellite. However, it will have little or no military value and will require about a decade or more of incremental improvement before it is capable of carrying out a basic military mission. A serious Soviet space laser effort will have to involve a manned platform to provide maintenance and repair services if the system is to have a reasonable "peacetime" operational lifetime. This raises the troubling possibility that the United States might have to contemplate destroying a Soviet manned spacecraft in order to protect American unmanned space systems from possible laser attack. In some ways the problem is similar to launching an attack against a Soviet air defense command post in the Middle East because it directs the shooting down of a U.S. reconnaissance drone.

There can be no doubt that the Soviet military is active in space. And, in the future, the Soviet Union will certainly increase its military investment in space. Its space hardware will improve, mission presence will increase, and the military utility of Soviet space-based assets will grow. Yet, the military significance of these Soviet activities cannot be evaluated in terms of simple-minded statistical indices or technical comparisons. Soviet military systems in space, whether serving in support missions or as weapons, are only one component of a large and diverse military force posture. The question should not be how much is the Soviet Union putting into space, but rather how much is it getting out of space?

NOTES

1. Office of Technology Assessment (OTA), *Salyut: Soviet Steps Towards a Permanent Human Presence in Space* (Washington, D.C.: Government Printing Office, 1983).
2. The main sources of reference used in this paper are Nicholas Johnson, *The Soviet Year In Space* (Colorado Springs, Colo.: Teledyne Brown Engineering, 1981 & 1982); U.S. Senate, Committee on Aeronautical and Space Sciences, *Soviet Space Programs 1971–1975* (Washington, D.C.: U.S. Government Printing Office, 1976); U.S. Senate, Committee on Commerce, Science, and Transportation, *Soviet Space Programs 1976–1980* (Washington, D.C.: U.S. Government Printing Office, 1982); James E. Oberg, *Red Star In Orbit* (N.Y.: Random House, 1981); "The Soviet Military Space Program", *International Defense Review* 15:2 (February 1982): 149–54; and Bhupendra Jasani, ed., *Outer Space—A New Dimension of the Arms Race* (London: Taylor and Francis, Ltd., 1982).

   Supplementary data collection was conducted by Dan Shephard.
3. Victor Yevsikov, *Reentry Technology and the Soviet Space Program* (Falls Church, Va.: Delphic Associates, 1982): 17.
4. In the case of civilian space launches it is reported that the Strategic Rocket Forces turns over control to civilian project managers ten minutes after a successful launch. U.S. Senate, Committee on Commerce, Science, and Transportation, p. 139.
5. This may be particularly true at Tyuratam, where a number of silo launchers for F-type boosters exist. See note 17.
6. The United States maintains an equivalent, if not superior, mission capability with only two launches per year. U.S. MPRSATs use either multiple film-return capsules or electronic data transmission.
7. Manuever capabilities allow Soviet MPRSATs to repeat ground tracks over areas of interest in order to follow developments over short periods. See Johnson; Jasani; and U.S. Senate, Committee on Aeronautical and Space Sciences, p. 457–78.

8. Stephen M. Meyer, "Anti-Satellite Weapons and Arms Control: Incentives and Disincentives from the Soviet and American Perspectives," *International Journal* 36:3 (1981): 460–84.

   The fact that the Soviets have not stressed the development of real-time space-based MPR systems leads one to doubt that they are thinking of such capabilities in the context of battle management/troop control. On theater reconnaissance see Victor Suvorov, *Inside the Soviet Army* (London: Macmillan Publishing, 1983).
9. Soviet military writers have talked about the need for an airborne armed reconnaissance capability to seek out and destroy U.S. mobile ICBMs.

   Soviet military planners have followed a parallel course in their planning for theater warfare. See Stephen M. Meyer, "Soviet Theater Nuclear Forces: Parts I and II," *Adelphi Papers* (London: International Institute for Strategic Studies, 1984).
10. The calculation is based on random failures distributed in time, according to a simple exponential distribution model. All other such calculations that follow are based on the same rule.
11. The resolution required in the theater, the quick response needed for the use of ESM, ECM, and ECCM, and the desires of operational commanders to regulate this aspect of "radio-electronic combat" imply that space-based ELINT would play a secondary role in theater warfare. See M.P. Atrazhev, V.A. Il'in, and N.P. Mar'in, *Bor'ba s radioelektronnymi sredstvami* (Moscow: Voenizdat, 1972); and V.A. Vartanesyan, *Radioelektronnaya razvedka* (Moscow: Voenizdat, 1975).
12. U.S. Senate. Committee on Commerce, Science, and Transportation, p. 430–45; Charles S. Sheldon II, "The Soviet Space Program in 1979," *Air Force Magazine* (March 1980): 88–93.
13. Johnson. The constellations are separated in a manner that allows for intersecting views by two satellites of missile launch areas.
14. Data from Johnson.
15. It is estimated that the resolution of the Soviet RORSAT radar is about 30 meters. R.T. Rees and C.P. Vick, "Soviet Nuclear Powered Satellites," *Journal of the British Interplanetary Society* 36:10 (October 1983): 457–60.
16. U.S. Senate, Committee on Aeronautics and Space Sciences, p. 398–429).
17. Some eighteen FOBS silos have been identified at Tyuratam. Under SALT II they were to be decommissioned. Currently, they are considered to be test launchers, though they it is not certain that they are empty.
18. Data from Johnson.

19. Contrary to common impressions, the Soviet military budget does not expand to accommodate every new project. Tradeoff patterns among and within the different branches of the Soviet Armed Forces are well documented. To some extent, the "phasing" of large military procurement programs is required in an economy where materials balances replace market forces.
20. M.M. Kir'yan, *Voenno-tekhnicheskiy progress i Vooruzhennye Sily SSSR* (Moscow: Voenizdat, 1982), Chapter 7; P.V. Morozov, *Bor'ba s vozdushno-kosmicheskimi tselyami* (Moscow: Voenizdat, 1967); and I.I. Anureev, *Oruzhie protivoraketnoy i protivokosmicheskoy oborony* (Moscow: Voenizdat, 1971).
21. Soviet weapons designers are put under tremendous pressure to develop a producible weapon, and many of the incentives in the system are oriented towards ensuring production rather than performance. Moreover, a capacity for modular improvement and incremental innovation is required. For example, one of the competing designs for the Moscow-Leningrad ABM system was rejected, in part, because it could not be upgraded in a piecemeal fashion. Based on an interview with A. Fedoseev, former chief designer of the Soviet magnetron laboratory.
22. Oberg; OTA; and V. Kotel'nikov, "The Orbits of Peace and Progress," *Aviation and Cosmonautics* (Moscow) 11 (November 1982): 22–23.

REFERENCES

Anureev, I.I. *Oruzhie protivoraketnoy i protivokosmicheskoy oborony.* Moscow: Voenizdat, 1971.

Atrazhev, M.P., Il'in, V.A., and Mar'in, N.P. *Bor'ba s radioelectronnymi sredstvami.* Moscow: Voenizdat, 1975.

Jasani, Bhupendra, ed. *Outer Space—A New Dimension of the Arms Race.* London: Taylor and Francis, Ltd., 1982.

Johnson, Nicholas L. *The Soviet Year in Space.* Colorado Springs, Colo.: Teledyne-Brown Engineering, 1981, 1982.

Kir'yan, M.M. *Voenno-tekhnicheskiy progress i Vooruzhennye Sily SSSR.* Moscow: Voenizdat, 1982.

Kotel'nikov, V. "The Orbits of Peace and Progress." *Aviation and Cosmonautics* (Moscow) 11 (November 1982).

Meyer, Stephen M. "Anti-Satellite Weapons and Arms Control: Incentives and Disincentives from the Soviet and American Perspectives." *International Journal* 36:3 (1981).

———. *Soviet Theater Nuclear Forces: Parts I and II.* Adelphi Papers Nos. 187 and 188. London: International Institute for Strategic Studies, 1984.

Morozov, P.V. *Bor'ba s vozdushno-kosmicheskimi tselyami.* Moscow: Voenizdat, 1967.

Oberg, James E. *Red Star in Orbit.* New York: Random House, 1981.

Rees, R.T., and C.P. Vick. "Soviet Nuclear Powered Satellites." *Journal of the British Interplanetary Society* 36:10 (October 1983).

Sheldon, Charles S., II. "The Soviet Space Program in 1979." *Air Force Magazine* (March 1980).

Suvorov, Victor. *Inside the Soviet Army.* London: Macmillan Publishing, 1983.

"The Soviet Space Program." *International Defense Review* 15:2 (February 1982).

U.S. Congress. Office of Technology Assessment. *Salyut: Soviet Steps Towards a Permanent Human Presence in Space.* Washington, D.C.: U.S. Government Printing Office, 1983.

———. Senate. Committee on Aeronautical and Space Sciences. *Soviet Space Programs, 1971–1975.* Washington, D.C.: GPO, 1976.

———. Senate. Committee on Commerce, Science, and Transportation. *Soviet Space Programs, 1976–1980.* Washington, D.C.: GPO, 1982.

Vartanesyan, V.A. *Radioelektronnaya razvedka.* Moscow: Voenizdat, 1975.

Yevsikov, Victor. *Reentry Technology and the Soviet Space Program.* Falls Church, Va.: Delphic Associates, 1982.

4 SPACE SYSTEM VULNERABILITIES AND COUNTERMEASURES

George F. Jelen

INTRODUCTION

Although the frontier of space offers many unique and attractive advantages, it is not without some countervailing disadvantages and limitations. This chapter discusses some of these limitations, stressing in particular some of the vulnerabilities of space systems and some of the ways in which these vulnerabilities can be reduced.

A given satellite is but one part of a complex space system whose proper functioning requires a launch system, command and control network, and communication links to connect the system's user, usually on the ground, with the orbiting satellite.[1] To disrupt the proper functioning of a typical space system, it is necessary only that any one of these four subsystems fail or be made to fail.

Secondly, many of the space systems—such as most navigation systems—depend upon more than one satellite in order to function. A particular navigational satellite system, for example, consists of eighteen satellites and requires simultaneous, line-of-sight contact with four of them in order for the system to operate. As increasing numbers of the eighteen satellites fail, the system is successively degraded. Even if only one of the eighteen should become inoperable, there are combinations of times and places on the earth's surface for which the system will no longer work. As more satellites fail, these areas and durations expand.

A final point to keep in mind is that space systems tend to be inherently fragile. Because they are very expensive to build and to launch, space systems are usually designed to the limit and possess little in the way of performance margin. This renders them inherently vulnerable either to unintentional mishap such as an occasionally uncooperative environment, internal failure, or human neglect, or to the direct malevolent action of an opponent. These vulnerabilities will be taken up in turn.

Space is not a completely benign environment. Satellites must withstand extreme temperature changes, and the near-vacuum of space offers little protection from external sources of radiation—particularly the sun. And even if the environment of space is, on balance, less hostile than that of earth, the satellite still has to get there. The shock and vibration stresses encountered during launch are severe, and special care must be given to a satellite's mechanical engineering to enable it to survive this battering. Also, satellites have to be able to endure without maintenance. Although the arrival of the space shuttle may soon change this, today's satellites are launched once and for all. They must, therefore, be reliable enough to survive for years without being touched. Subsystem redundancy, in which virtually all black boxes are duplicated and cross-connected, is standard practice throughout the space industry. Nevertheless, in spite of this redundancy, satellites do fail. Components do not live forever, even in space, and even duplicated subsystems eventually malfunction or cease working altogether. However, enough experience has been accumulated with spacecraft designs that satellites, once in orbit, are likely to last ten years or more. Eventually, they run out of the fuel required to maintain them in a particular orbit, and they are turned off lest their transmissions constitute a source of noise for other satellites. Absent hostile action, then, space systems have become quite dependable. It is not surprising, therefore, that the United States, as well as many other countries, have grown to depend on space systems for civil and military functions. It should also not be surprising that, in light of this dependence, satellites have become high-priority targets for hostile action by military opponents.

Satellites make likely targets for yet another reason: They are hard to defend. It is nearly impossible to hide them; they travel in very predictable trajectories; and the near-vacuum in which they operate offers little shielding against various forms of energy that might be directed against them (these properties of space are discussed at greater length in Chapter 6).[2]

HOSTILE ACTION

There are many ways to "kill" a space system, ranging from direct, brute force actions like physical destruction to more subtle measures, like conning the system into taking destructive action against itself or failing to take some action vital to its continued health.[3] The first and most obvious strategy is to attack the satellite itself, with the intention of inflicting irreparable physical damage. Such irreparable damage may be inflicted in a number of ways: by direct collision, by fragmentation, by means of directed energy beams, or by means of electromagnetic pulse (EMP).

The major antisatellite (ASAT) system under development by the United States is designed to destroy its satellite target by kinetic energy; that is, by actually striking the target at high velocity.[4] It is to be carried aloft and launched by an F-15 fighter. A ground station tracks both the aircraft and its target until the ASAT is launched into what is known as a high-altitude zoom climb. The ASAT is a three-stage vehicle; the first two stages are inertially guided and the third stage, called the "miniature homing vehicle," employs infrared homing.[5]

U.S. antisatellite weapons are to be commanded from the Space Defense Operations Center located in the underground headquarters of the North American Aerospace Defense Command deep within Cheyenne Mountain near Colorado Springs. The U.S. system is not expected to be ready for operational use until 1987.[6]

The Soviet ASAT system, already considered operational by the Department of Defense, is satellite- rather than aircraft-based. The Soviet ASAT is guided from its orbiting position into close proximity to its target and then exploded. Fragments from the explosion destroy the target.[7] Three different attack modes have been observed in Soviet ASAT tests. In the first, the ASAT weapon is launched into an elliptical orbit that intersects with the orbit of the target. In this mode, the closing velocity between weapon and target tends to be moderately high. In the second mode, the ASAT is launched into an orbit that is very similar to the orbit of the target. The closing velocity in this mode is much slower. Finally, in the "pop-up" mode the ASAT does not complete a single revolution around the earth; the target is engaged during the initial orbit.[8]

The Soviet system appears to be limited in capability. Although the Soviets have conducted at least twenty tests of the system since 1968, they do not appear to have settled on a method of target acquisition, having been noted using both radar and infrared. The system has only been tested at inclinations between 62 and 66 degrees and at altitudes between 550 and 2,000 kilometers. To what extent these reflect the actual limits of the system is unknown. Finally, the fact that a very large liquid-fueled vehicle is used to launch the Soviet ASAT probably imposes some constraints on how rapidly the system can be launched. Even with these limitations, however, it represents a serious threat to certain U.S. satellite systems—the Navy's ocean reconnaissance satellite, for example.[9]

Another form of hostile action might involve space mines—in essence, orbiting bombs that can be detonated by command from the ground. They would be particularly useful against satellites in geosynchronous orbit. In such an orbit, a space mine could be positioned near its target where it could remain in a dormant state for years. It could even be disguised as something else—a communications satellite, for example. Then, at the time chosen for it to attack, an internal homing device would be turned on, causing it to lock on to the target. The weapon could then maneuver within lethal range and explode.[10]

Beam weapons (often called "directed energy" weapons) are a category embracing several different weapon technologies including lasers, particle beams and heat rays. Although all three technologies show some promise of eventually producing effective weapons, such weapons are at least several years away. Since the technology, feasibility, and utility of lasers and particle beams are treated extensively in Chapter 6, they receive only brief treatment here.

Laser light is coherent; that is, all of the light is at virtually the same wavelength, moving in one direction, and oscillating in phase. The waves in a laser, therefore, reinforce one another, and the resultant beam is straight, narrow, and very sharply focused over considerable distances. Very short pulses of laser energy can reach extremely high intensities and instantly create local pressures as high as a million atmospheres. Such pressures have already proven highly effective in punching holes in solid matter, without affecting the surrounding material.[11] Theoretically, a laser could shift its focus from one target to another very quickly, thus permitting an extremely rapid rate of fire. Since atmospheric particles diffuse light, the less atmosphere a laser has to contend with, the better it operates. Although all of the necessary theory is well understood, there remain significant practical difficulties in producing the required power generators and the aiming devices. Overcoming these practical problems is estimated to require billions of dollars and several additional years of development.[12]

Nevertheless, both the United States and the Soviet Union are busily engaged in the development of laser weapons that could disable orbiting satellites. The Soviet Union may currently be slightly ahead. According to a *New York Times* article, "some analysts estimate that the Russians may send a crude weapon into orbit in one to five years."[13] The U.S. goal appears to be more modest. Although expenditures have been increasing sharply, the United States is not expected to produce a laser weapon for five to ten years. Recent initiatives by the Reagan administration, however, seem to be aimed at advancing that date. There are still many who even question the ultimate utility of a laser weapon. The criticism is that a ground-based laser weapon would only threaten satellites in low orbit; a space-based laser would itself be vulnerable to attack by less expensive and less complex weapons systems; and both may be less effective than other more conventional ASAT weapons.[14]

Particle beam weapons make use of streams of charged or neutral atomic or subatomic particles, such as protons, neutrons, and the like. Neutral particle beams, like lasers, are diffused by the earth's atmosphere. Charged particle beams, on the other hand, generally perform better within the atmosphere because they tend to spread out and become less focused unless they can interact physically with the surrounding air, which permits particle replacement within the beam. Therefore, against a spaceborne target, the most effective placement of a charged particle beam weapon would appear to be on the ground (for a discussion of space-based neutral beams, see Chapter 6). Particle beams are able to penetrate more deeply into their target than lasers and thus possess a wider range of kill mechanisms.[15]

Particle beam weapons appear to be even further into the future than lasers. Nevertheless, both the United States and the Soviet Union are actively researching their potential.[16]

The third type of beam weapon is the heat ray. The use of a heat ray as a weapon is not a new idea. Archimedes is generally credited with the construction of a "burning mirror" during the siege of his native city, Syracuse, in 212 B.C. By means of this device, Roman ships were reportedly set on fire as soon as they came within bow shot.[17] A heat ray weapon, theoretically, could be developed to harness the rays of the sun or could employ its own indigenous heat source—such as might be generated by a small, confined thermonuclear reaction. The difficulty in "confining" a nuclear reaction lies in coping with temperatures that range from 45 to 400 million degrees Kelvin. No known material can stand up to this heat. It may eventually become possible to confine the effects of the nuclear reaction magnetically, but this lies well beyond present capability.[18] Thus, however our imaginations might be stirred by the contemplation of such weapons, they remain, in the opinion of nearly all, appropriate subject matter only for science fiction.

The final form of overt, direct attack is by means of an electromagnetic pulse (EMP), an intense burst of energy triggered by a nuclear explosion. It results from the collision of gamma rays produced by the nuclear blast with electrons in the air molecules of the upper atmosphere. For a 10-megaton burst at an altitude of 400 kilometers, this region of collision is reportedly about 3,000 kilometers in diameter and 10 kilometers thick. The electrons thus scattered and accelerated by these collisions are called Compton electrons (after their discoverer Arthur H. Compton). When Compton electrons encounter the earth's magnetic field they are deflected and produce a transverse electric current. This current, in turn, sets up the electromagnetic pulses, which propagate downward toward the earth. The peak electrical fields produced in this way could be as high as 50,000 volts per meter.[19]

An EMP attack could be launched against either the ground or the space segment of a satellite system. Any electrical conductor within the large area of EMP effects will act as an antenna to pick up the electromagnetic pulse. EMPs can affect an area on the ground of 500 to 800 kilometers or even more in radius, depending upon the height and yield of the burst, and even a relatively small explosion of about two megatons just outside the earth's upper atmosphere (from 80 to 120 kilometers altitude) would damage an unprotected satellite in geosynchronous orbit. By increasing the size of the bomb, this kill radius can easily be increased.[20]

The vulnerability of different components to EMP effects varies widely. First of all, EMP vulnerability increases with electronic complexity. On average, it takes approximately one hundred times as much EMP energy to damage vacuum tubes as it does transistors, and one hundred times as much energy to damage transistors as it does integrated circuits. Also, similar components of the same technology can vary by several orders of magnitude depending upon the manufacturer, and even within a single batch of components from the same manufacturer, EMP sensitivity can vary as much as two orders of magnitude.[21] Accurate prediction of the EMP susceptibility of a given system, therefore, is quite difficult.

Nuclear explosions can have other, longer term, deleterious effects on electronic components besides the instant EMP effect. They can also result in trapped electrons that can reduce a spacecraft's life by creating noise in its sensors and by slowly degrading its electronic components and solar cells.[22] And, of course, if close enough to the intended target, the radiation flux from a nuclear explosion can destroy a satellite. During the 1960s, the United States actually deployed an antisatellite system equipped with nuclear warheads; the system was dismantled in 1975.[23]

PHYSICAL COUNTERMEASURES

Although it is true that for every measure there is a countermeasure, it is also true that if an opponent is willing to expend sufficient resources, any space system can be defeated. On the other hand, most can be protected against a "cheap shot."[24] Such appears to be the philosophy that underlies current American space policy.

Among the options available for rendering satellite systems less vulnerable are physical hardening, making the satellites more difficult to detect or track, evasive maneuvers, higher orbits, and an increased degree of self-sufficiency or autonomy.

"Hardening" is a very broad term that is used to cover a large number of specific actions—all aimed at making a given system less vulnerable to physical attack. Because a large number of satellites are controlled by a much smaller number of ground stations, a very high priority of any hardening effort must go to the ground segment. Fixed ground stations are particularly vulnerable to a direct nuclear or conventional attack because their antennas must always remain exposed. The U.S. Air Force recognizes this latent vulnerability.[25]

Among the goals of any effort to harden the ground segment would be preventing EMP coupling from high-altitude nuclear bursts and minimizing vulnerability to sabotage and terrorist attack. Methods of EMP protection are well known. Vulnerable circuits must be protected against direct EMP radiation as well as against EMP-induced surges from connected conductors. The electromagnetic waves can be countered by encasing equipment or whole facilities in metal Faraday shields or simply by burying them deeply enough. The conductor-carried pulses can be isolated from equipment by surge arrestors or filters, somewhat similar to those used to protect against lightning but designed for the much faster rise-times associated with EMP-induced currents.

The problem with these countermeasures is cost. Shielding for entire power or communications grids, which could be accomplished either by burying the entire grid or by shielding every piece of equipment within it, is economically unfeasible. According to one source, "the only feasible alternative is the shielding of local facilities," while "simultaneously isolating them from the surrounding grid by surge arrestors." Such measures, however—particularly when applied to existing facilities—can raise the price of the facility from 10 to 100 percent.[26]

To protect against sabotage or an attack by terrorists, care must be given to the location of the site and to its physical security. Even more effective is to reduce dependence upon any one location by employing redundant and backup systems. As a step in this direction, a satellite operations complex (SOC) has been proposed as a backup to the Satellite Test Center in Sunnyvale, California—the present nerve center for most military space systems. The SOC and a planned shuttle operations and planning complex (SOPC) are to be co-located within a facility called the Consolidated Space Operations Center, near Colorado Springs.[27]

Satellites can also be hardened. Nuclear effects, EMP, and trapped electron effects can be mitigated through the use of specially designed electronic components, shielding sensitive parts, and generally designing the system to withstand greater amounts of radiation. "All of these concepts," says Air Force Colonel Robert B. Giffen, "are within the state-of-the-art and, when employed, reduce the probability of the enemy's achieving multiple kills with a single nuclear blast."[28]

If the threat is from an orbiting interceptor (ASAT), other countermeasures are available. As Giffen observes, "The key to defeating an orbital interceptor is first to know that he's coming. Once you know he's coming, then the object is to defeat his terminal guidance system, which will be radar, infrared, or optical."[29] Radar signals can be jammed and infrared systems can be spoofed through the use of decoy heat sources; decoys, in the form of replica satellites or even light images, can be used to confuse optical sensors; and signature reduction, often referred to as "stealth technology," can be employed. This latter technology aims at minimizing the electromagnetic emissions and reflections by which these various sensors spot, identify, and track their targets. Means include special paints and coatings that do not reflect radar waves very well as well as careful attention to shape and surface material.[30] This technology, however, though applied successfully to bombers and missiles, is difficult to apply to satellites. Solar panels, for example, present a particular challenge.[31]

Orbital interceptors can also be countered simply by moving out of their way. Of course, to do this requires both advance knowledge and sufficient time to carry out the necessary maneuver. Therefore, in general, the higher the orbit, the more useful a maneuvering capability is. The minimum time of flight to geosynchronous altitude, for example, is from three to six hours. Even a small maneuver a couple of hours prior to interception can place the target well out of range.[32] The U.S. Air Force seems to place a great deal of stock in satellite maneuverability.[33]

There are two disadvantages to this form of countermeasure, however: Payload weight is sacrificed for additional fuel, and many satellites (those requiring precise pointing, for example) cannot maneuver and perform their missions simultaneously. Maneuvering also offers little protection against beam weapons or against space mines located abreast of their targets.[34]

The impact of individual satellite vulnerability can be reduced through the use of orbiting spares. Spare satellites can be stored in high orbits and then brought down to replace a damaged or destroyed satellite, or they can be inserted into their normal operating orbit but left silent or turned off until needed.[35] Storage in the higher orbit offers added safety but requires extra fuel to place and later retrieve the satellites.

To remain effective, any countermeasure must be as flexible as the system it is attempting to counter. If the opponent's system can change frequencies in the presence of jamming, for example, the countermeasure system must possess this same capability. And, of course, once a satellite is in orbit, its countermeasure capability tends to be fixed. As Giffen observes, "Retrofit of an orbiting satellite is difficult!"[36]

A space system's overall vulnerability to disruption can be reduced if the role of its most vulnerable components can be reduced—particularly the role of the ground segment. There is currently within the U.S. Air Force a vigorous effort toward what it calls "satellite autonomy." The Air Force, according to one account, defines satellite autonomy as "the capability to perform selected functions without ground support at specified levels of conflict for a given period of time." The Air Force has established for itself the goal of six-months autonomous operation for all military satellites. It has created a formal Satellite Autonomy Program within the Space Division and has gone so far as to quantify levels of autonomy from zero (total dependence) to ten (total independence). Six-months autonomous operation is a "level five" capability.[37]

COMMUNICATIONS LINKS

Short of direct, physical attack, the most damaging actions one can take against a satellite system are those involving its communications paths. The utility of almost all satellite systems depends on their ability to receive and to send messages.

There can be as many as four distinct communications links or paths associated with a space system: a ground-to-satellite link, or uplink; a satellite-to-ground link, or downlink; a ground-to-ground link through the satellite, or throughlink; and a satellite-to-satellite link, or crosslink. Any one of these links may contain several channels of data, but not all links are present in all space systems. The throughlink, for example, is generally present only on communications satellites, and only a few space systems make use of a crosslink between satellites, although this use is increasing. We will discuss each of these links in turn because each has its own peculiar set of vulnerabilities.

The uplink, which carries messages from the ground to the satellite, serves primarily as the command channel, passing orders from command and control centers to the satellite. It can also pass along information important to a satellite's mission, which can be stored in the satellite's memory—for example, specific targeting information for a reconnaissance satellite's next orbit.

All modern space systems contain a command channel—part of the uplink. It is one of the single most important communications links in any system. Without it, most present day systems would not long function. Giffen, for example, states:

> Many satellites require constant monitoring to turn systems on and off, maneuver, maintain stable pointing attitudes, function properly in the earth's shadow, keep proper spin ratios, etc. The command and control demands may be so high that the satellites would fail catastrophically within a few hours without help from ground controllers.[38]

Without regular commands or corrections from the ground, most satellites will drift, due to the gravitational attraction of the moon and the sun. Satellites in synchronous orbit, for example, must make periodic east-west position corrections or they will drift in longitude. And without occasional north-south corrections, the orbit inclination will drift as well.[39]

The second of a space system's communications links is that which originates on board the satellite and terminates on the ground—the downlink. The downlink is used for three principal purposes: to transmit information that the satellite has collected about something else (as in a reconnaissance satellite), to transmit information about the satellite itself, and to facilitate the tracking of the satellite.

The information a satellite transmits about itself is called telemetry, from the Greek *tele*, meaning distant, and *metron*, to measure. As data for a telemetry signal, physical phenomena are expressed in terms of voltage analogs. Temperature, power levels, magnetic fields, fuel consumption, particles, and so on are sensed by instruments that translate these measurements into varying electronic voltages. These voltage analogs are then converted into digits or numbers and transmitted as a stream of binary digits (bits).[40]

A satellite can be tracked more readily by way of a beacon, another downlink signal. A beacon emits an easily recognizable signal on a constant frequency or set of frequencies. Communications satellites often insert a beacon in every one of their output channels, just above or just below the frequency band used for communications. By deliberately using an extremely stable frequency source for the beacon signal, the receiving station can track frequency drifts in the communications channel.[41]

Throughlinks—communications paths through a satellite—are of two basic types. The two differ from each other in the amount of signal processing that is done on board the satellite. In the simpler type of throughlink, the communications signal to the satellite is not even demodulated. It is simply received, translated in frequency, amplified, and retransmitted. In this case, the satellite acts very much like a microwave repeater station.

The more complicated system is one in which the satellite processes the signal before retransmitting it. The processing might involve switching, demodulation, or both. Switching permits input channels to be connected by ground commands to output channels in different ways, increasing flexibility. On-board demodulation of the input signal can improve the performance of the link because it involves the regeneration of the signal when it reaches the satellite—after only half of its total journey.[42]

The last of the communications paths found in space systems is the crosslink. This path connects one orbiting satellite with another. If one satellite is dependent upon another for the relaying of its information to its ultimate destination, then the link between these two satellites is a critical one. Nevertheless, the crosslink is not usually an attractive target for hostile action, because at the extremely high frequencies that are typically employed the available channel bandwidths are quite high and specific frequencies can be selected that are greatly attenuated by the earth's atmosphere. Higher frequencies also result in narrower or more focused beams, which impose geographical constraints on the would-be attacker. The relay satellite itself, however, might be a high-priority target for the kinds of direct physical attacks already discussed.

THREATS TO COMMUNICATIONS LINKS

As with physical threats, there are several things that can disrupt communications without any help from an opponent. Any communications signal is susceptible to interference and atmospheric fades, and, of course, the communications equipment itself can fail. However, all of these phenomena are reasonably well understood, and present-day space systems can be and are designed to cope adequately with all of them. We are more concerned, here, with hostile actions an opponent can take deliberately.

In a general sense, there are three basic strategies that an opponent might pursue against each of the four communications links. First, he might attempt to make use of the link himself to his own advantage.

Second, he might attempt to deny the use of the link to the legitimate user. Or, third, he might inject his own signal into the link with the intention of deceiving either the system itself or a user of the system.

The means by which an opponent might attempt to carry out these strategies include collection, jamming, and spoofing. Collection consists of passive interception and perhaps recording of the target signal. Jamming involves the deliberate transmission of a competing signal at the same frequency as the target signal with the intent of interfering with its reception. Spoofing consists of the deliberate transmission of a signal that looks very much like the true target signal with the intent, not of interfering, but of deceiving the legitimate user. Each of the four links can be examined as to its respective susceptibility to these various strategies and means.

Command links probably present the widest range of choices for the would-be opponent. He could temporarily deny or limit the use of the satellite by jamming the link. If this is done at a particularly critical time, such as during launch, the effect could even be catastrophic and permanent. Or, he could intercept the signal in order to gain important information regarding the satellite's purpose and the precise way it is controlled. This information would be necessary if he wanted to take control sometime in the future, an action that falls into the third category.

But by far the greatest threat to a command link, in terms of potential mischief, is spoofing. If an opponent is able to generate his own set of commands and get the satellite to act on them, there is almost no limit to what he can do within the bounds of the satellite's capability. Also, if done carefully, spoofing is virtually undetectable and certainly almost impossible to prove.[43]

Fortunately, for each of these actions, there is a countermeasure. Against the threat of jamming, one can employ some form of frequency agility or band spreading; against the threat of interception and recovery, there is encryption; and against the threat of spoofing, there exist digital signatures—means of validating the authenticity of a transmission. These techniques will be discussed at greater length later in the chapter.

Geometry makes downlinks much harder to attack than uplinks, this in spite of the fact that the size of the downlink transmitting antenna as well as output power have to be limited. More than compensating for this latter limitation is the fact that the receiving antenna on the ground can be quite large and, as a consequence, can be highly directional, focused at the satellite itself. Collection of the downlink signal is a relatively easy matter, but to spoof it or to actively jam it almost requires that the opponent situate himself within the beam of the receiving antenna—that is, between that antenna and the satellite—usually no small feat.

Throughlinks can be targeted either on the way up or on the way down. To an opponent, the upward path of a throughlink is very much like an uplink and the downward path much like a downlink. The earth-to-satellite path is easier to jam or spoof but can be collected only by a receiving station very near the transmitter or from a mobile receiving station such as an airplane or satellite that is within the beam of the upward path. The satellite-to-earth path, on the other hand, is easier to collect but harder to jam or spoof. However, since on a throughlink the information on the upward path and the downward path is the same, the opponent has a choice as to which of the two paths to target, and obviously he will choose the path that optimally accommodates his selected strategy. If he has decided to collect, he will target the downward path. If his choice is to jam or spoof, he will concentrate on the upward path.

Crosslinks present the most formidable task to the opponent, and therein lies much of their value. A successful attack against a crosslink would almost certainly have to be conducted from an orbiting position, somewhere along the extended line defined by the locations of the two communicating satellites. To carry out all three hostile actions of collection, jamming, and spoofing would require a position *between* the two satellites. Collection could also be accomplished from behind the receiving satellite but not from behind the transmitting satellite. Jamming or spoofing would not be effective from behind the receiving satellite but could be done from behind the transmitting satellite. If the crosslink were a two-way path and if the opponent wished to attack both paths, he would either require two positions or would again be forced to align himself between the two communicating satellites.

There is a tendency on the part of some people to belittle some of the more sophisticated or exotic threats. One hears words to the effect of "Oh sure it's possible, but nobody would really do those things." The following quotation from a book about Army communications during the Vietnam War should help still such skeptics:

> There are numerous instances on record of the enemy jamming radio frequencies and sending false messages. These bogus transmissions used

imitation to try to turn fires or forces to an area chosen by the enemy. In one case the enemy tapped the internal telephone lines of a defensive base and diverted reserve forces from the area where he attacked.[44]

We might recall, moreover, that the Vietnam War was conducted between ten and twenty years ago against a relatively unsophisticated opponent. The Soviet Union, a sophisticated opponent, is known to give high priority to all forms of anticommunications actions, including deception.[45]

COMMUNICATIONS COUNTERMEASURES

Several techniques are available to protect communications links from would-be attackers or opponents. Different techniques are effective against different attack strategies. Against collection, available countermeasures include highly directional beams, low power, encryption, and advanced low probability of intercept (LPI) techniques. Directional beams, low power, and LPI techniques all aim at making the signal harder for an opponent to intercept. Encryption makes the signal no harder to intercept but, if done properly, renders the signal unintelligible to all but the intended recipients.

More directional beams, or narrower beamwidths, can be achieved by increasing the size of the transmitting antenna, by increasing the frequency of transmission, or both. Normally, there are practical constraints on both of these measures, however. If the signal must pass through the atmosphere, many frequencies above 10 Gigahertz (ten billion cycles per second) have to be avoided because the atmospheric attentuation due to water vapor and oxygen becomes significant.[46] The maximum thrust provided by the launch vehicle imposes constraints on any spacecraft's size and weight that in turn restricts antenna size on board the satellite. In general, it is probably fair to say that neither measure affords much intercept protection against a determined opponent.

Another way to make collection more difficult is to reduce the transmitted power. Obviously, this can only be done on a fairly limited basis since reducing the power will also adversely affect reception by the intended party. Also, as discussed below, reducing power increases the susceptibility to jamming and spoofing.

The standard way that military satellites, at least, thwart collection is by means of encryption. The word "encryption" or "encipherment" refers to the process by which the "raw" message (called plaintext) is transformed into disguised form (called a cryptogram or cipher text), intended to render the contents of the original message secret. The reverse process of transforming ciphertext back to plaintext is called decipherment or decryption. Both processes are controlled by a cryptographic key, which must be shared between sender and all intended recipients. If the encryption process is properly designed, an opponent lacking the correct key, even if he knows the basic encryption scheme, is forced to try all possible keys in order to decipher the message. The process that an opponent uses to "break" the cipher is called "cryptanalysis."[47]

In some specialized cases, denying the content of a transmission to one's opponent is not enough. One may wish to keep secret even the existence of the signal, or to conceal its source. In such cases, a more sophisticated technique, known as low probability of intercept (LPI), is employed. There are two basic methods of achieving LPI. One is by means of extremely short or "burst" transmissions. The second is through the use of spread spectrum techniques—similar to those used to defeat jamming.

In burst transmissions, the digital message is stored or buffered, compressed in time, and then transmitted as a very short spurt of high-speed data.[48] As the information-carrying modulation is thus sped up in time, the bandwidth of the signal is correspondingly increased. This combination of a very short signal and high bandwidth makes the signal very difficult to detect and, therefore, to collect.

Spread spectrum LPI signals employ such a broad band of frequencies that the LPI signal would appear to an ordinary receiver as nothing more than wideband noise.[49] The two most common methods of achieving band or spectrum spreading are frequency hopping (FH) and direct-sequence pseudonoise (PN).

In FH systems, the radio frequency (RF) of the transmitted signal, instead of remaining constant as with ordinary radio transmissions, is shifted or "hopped" from one frequency to another. The pattern of hopping is ordinarily done according to some code sequence. In typical FH systems, the data rate is faster than the hop rate. Therefore, each hop will contain several data pulses. Because of its relatively slow rate of hopping, this method of band spreading is not normally employed for LPI purposes but can be an effective means of countering the effects of jamming.

In direct-sequence PN systems, a pseudonoise sequence is used to alter or shift the phase of the RF carrier according to some pattern.

The effect of this phase shifting is to widen the bandwidth of the transmitted signal and to make it appear more noiselike. The phase shifting is quite rapid. The rate that the phase is shifted, called the chip rate, is many times faster than the data rate so that each "bit" of information is actually spread over a wide frequency band or spectrum. This is the method normally used in LPI systems.[50]

There are two other methods of spectrum spreading. In the first, known as "time hopping," the RF signal is dispersed in time with some code-determined gap between signal elements. In the second, called "chirp," the transmitted radio frequency is swept or slid at a constant rate either up or down. A given system might use an up-sweep to transmit a binary "one" and a down-sweep to transmit a "zero".[51] However, neither time-hopping nor chirp systems are commonly used.

As was pointed out earlier, if the opponent's strategy is to deny the use of the signal to its owner, he will likely employ jamming. The simplest countermeasure to jamming is simply to increase the transmitted power. In the absence of jamming, successful communication depends upon some minimum ratio of the intended signal to all other received signals (which to the receiver constitute noise). In fact, signal-to-noise ratio is a standard figure of merit by which communications links are measured. As seen by the receiver, jamming is but another form of noise. Therefore, in the presence of jamming, successful communication depends upon a minimum ratio of the intended signal to the jamming signal plus all other sources of noise. Obviously, a straightforward way of increasing this ratio is by raising the power of the transmitted signal, assuming that more power is available and the increase does not saturate the receiver. Receiver saturation causes distortion, which is, in effect, but another form of jamming—in this case, self-jamming.

Frequency agility or band spreading (the LPI techniques discussed above) can also be used against jamming. For anti-jamming (AJ) purposes, both frequency hopping and pseudonoise methods are commonly used. Both methods are quite effective against unsophisticated jammers.

A jammer's success depends upon its ability to overwhelm the intended signal by placing more power at the receiver at a given instant of time than the legitimate transmitter. There are, however, very definite constraints. The amount of power a jammer can deliver to an opponent's receiver is constrained by energy requirements, size, and geography. Because of these limitations, a jammer's design must trade off signal power at any given frequency against the range of frequencies it can operate against. The wider or more diverse the band, the lower the average power it is capable of producing.

The key to success for either anti-jamming technique is the superior knowledge of the intended receiver. It "knows" where the legitimate transmitter will be concentrating its power in the next instant and can tune its filters accordingly. The jammer must try to anticipate the next frequency or set of phases just as the legitimate receiver does, by attempting to follow the transmitted frequency through a combination of wideband collection and rapid retuning or by attempting to decode the PN sequence. In either case, we are talking about a sophisticated jammer.

The first technique, which makes use of what is called a follower jammer, can only hope to cope with an FH system. It would not work against a PN system at all. Its effectiveness against an FH system depends upon the hop rate of the AJ system and the response time of the jammer. In such a contest, the AJ system would appear to have the distinct advantage, but perhaps a more accurate statement would be to say that the advantage falls to whichever of the two systems was designed last, since for every measure there is a countermeasure and vice versa.

The effectiveness of the anticipating or "intelligent" jammer, as it is usually called, depends upon the jammer's ability to recover the coding sequence. If a highly repetitive code is used, an intelligent jammer might well succeed, but if a more complex code is employed, even an intelligent jammer is likely to have great difficulty.[52]

Transmitting at much higher frequencies (above 10 GHz) also magnifies the task of the would-be jammer. Raising the frequency narrows the transmitted beam, which imposes much more severe geographic constraints on the jammer since the farther it is outside the beam, the less effective it is. Using such frequencies also has some advantage in avoiding the type of jamming that results from high altitude nuclear blasts. These frequencies reduce the duration of the nuclear absorption and scintillation effects, which block or distort transmission, from minutes to seconds.[53]

Finally, as a way of contending with spoofing, one can employ authentication techniques, usually called "digital signatures" because they represent an attempt to do for a digital transmission what a written signature does for a document or banking transaction. Authentication can be of the sender, of the message, or of both. If the digital signature is intended to authenticate the sender, it must make use of information or employ a process known or possessed only by the sender. It should not be possible for anyone else to duplicate nor can it be derivable from the collection and analysis of many messages from that particular sender. If the digital signature is to authenticate the message, it must change when the message changes. A different message must generate a different signature. And if it is to authenticate both the sender and the message, it must possess both properties simultaneously, so that either the same message sent by a different party or different messages from the same party would have different signatures. In a broad sense, both properties are achieved in the same general way. They both require the addition of redundancy to the intended message, not unlike the way in which transmission errors or "garbles" can be detected and corrected.[54]

Authentication as a means of countering a spoofing threat is particularly important on command uplinks. Without some form of command authentication, an opponent could transmit his own command to a satellite and, in effect, assume control of it.

Some have offered encryption as a means of countering the spoofing threat.[55] Although helpful, depending upon the means of encryption, this may or may not be totally effective. For example, some encryption schemes might be vulnerable to a record and playback threat, in which a valid encrypted command is intercepted by the opponent, recorded, and then retransmitted to the satellite later and however often the opponent might desire. How damaging this might be to a particular system would depend upon what the command was and, perhaps, when it was transmitted. Presumably, the opponent would not know the effect, but the action is such a simple one and it is hard to imagine too many circumstances in which it would do any good. Encryption, however, can be combined with authentication so that, in addition to establishing the legitimacy of the message, its contents are protected as well.

CURRENT EFFORTS

By their actions, both the United States and the Soviet Union have evidenced concern over the potential vulnerability of their space systems. Both countries have taken steps to enhance the survivability of their systems. U.S. military satellites routinely encrypt uplink and downlink signals and authenticate uplink commands. And the United States has identified two major current programs, the MILSTAR communications satellite system and the NAVSTAR Global Positioning System, as "pivotal programs" in the Air Force's survivability effort. As the commander of the Air Force Space Division puts it, "We can only afford a certain level of survivability, and these two programs are the most important from a survivability point of view."[56]

Both satellites are making use of the extremely high frequency (EHF) band and of crosslinks to enhance their communications survivability. MILSTAR uses EHF for all its links; NAVSTAR uses it for telemetry. A crosslink capability has been added to both satellites to permit ground controllers to communicate with the satellites through another satellite serving as a relay.[57]

For its part, the Soviet Union now encrypts "virtually all data" from its spacecraft, whereas, in the past, "transmissions from manned and scientific spacecraft were uncoded."[58]

THE ADVANTAGE OF THE LAST MOVE

We have seen that for every attack strategy that an opponent can employ there is a suitable defense that counters it. Even so, the attacking or offensive side always retains a most important advantage. In the series of moves and countermoves that characterize the continual battle for technological dominance between offense and defense, it is important to keep in mind that the offense always has the last move. It is the offensive side that controls the timing of the actual engagement. The offense, therefore, can optimize both the timing and the attack strategy against a generally known defense. Although an adequate defense may well be feasible, there will be no time, in actual conflict, for the defensive side to field the appropriate countermeasure once the engagement has begun.

NOTES

1. Robert B. Giffen, *U.S. Space System Survivability: Strategic Alternatives for the 1990s,* National Security Affairs Monograph Series 82-A, (Fort Lesley J. McNair, Washington, D.C.: National Defense University Press, 1982), pp. 15–16.

2. Daniel O. Graham, *High Frontier: A New National Strategy* (Washington: High Frontier, Inc., 1982), p. 27.
3. Giffen, pp. 25–31.
4. Graham Warwick, "The Military Use of Space," *Flight International* (4 December 1982): 1638.
5. Ibid., pp. 1637–38.
6. Craig Covault, "Space Defense Organization Advances," *Aviation Week & Space Technology* (8 February 1982): 21; and Philip M. Boffey, "Pressures are Increasing For Arms Race in Space," *New York Times,* 18 October 1982.
7. Ibid. See also Craig Covault, "Universe Red," *Omni* (August 1980): 53.
8. Warwick, p. 1637.
9. Thomas Karas, *The New High Ground: Systems and Weapons of Space Age War* (New York: Simon and Schuster, 1983), pp. 149–50.
10. Giffen, p. 28, and Karas, p. 161.
11. John W. Macvey, *Space Weapons/Space War* (New York: Stein and Day, 1977), p. 129.
12. James Canan, *War In Space* (New York: Harper & Row, 1982) pp. 147–48.
13. Boffey.
14. Ibid.
15. Donald M. Snow, "Would Laser Weapons Make the World Safer?" *Christian Science Monitor,* 2 March 1983; and *Aviation Week & Space Technology* (7 February 1983): 16.
16. Snow, p. 22.
17. Macvey, p. 81.
18. Ibid., p. 84.
19. Eric J. Lemer, "Electromagnetic Pulses: Potential Crippler," *IEEE Spectrum* (May 1981): 41.
20. Ibid., p. 42; and William J. Broad, "Frying All Our Electrons With a Nuke Over Omaha," *Washington Post,* 5 December 1982.
21. Lemer, p. 43.
22. Giffen, p. 37.
23. Boffey.
24. Ibid.
25. Interview with Lt. Gen. Richard C. Henry, Commander, USAF Space Division. In Edward I. Gjermundsen, "Satellite Autonomy—Assuring a Sustaining C3I Capabilility," *Signal* (February 1983): 24.
26. Lemer, p. 44; and Giffen, p. 35.
27. Ibid., p. 36; and Merritt E. Jones, "DOD's Control Center for Space Shuttle—SOPC," *Signal* (February 1983): 45.
28. Ibid., p. 37.
29. Ibid., p. 36.

30. Karas, p. 166.
31. P.J. Friel, "U.S. Space-Based Reconnaissance Communications and Command and Control Systems" (Paper presented at the Eleventh Annual Conference on *International Security Dimensions of Space,* sponsored by the International Security Studies Program, the Fletcher School of Law and Diplomacy, Tufts University, 27–29 April 1982), p. 24.
32. Giffen, p. 38.
33. Richard Halloran, "U.S. Plans Big Spending Increase For Military Operations in Space," *New York Times,* 17 October 1982.
34. Giffen, p. 39.
35. Karas, p. 167.
36. Giffen, p. 36.
37. Gjermundsen, p. 24.
38. Giffen, p. 32.
39. James J. Spilker, Jr., *Digital Communications by Satellite* (Englewood Cliffs, N.J.: Prentice Hall, Inc., 1977), p. 137.
40. Alfred Bester, *The Life and Death of a Satellite* (Boston: Little, Brown, and Company, 1966), pp. 91–92.
41. Spilker, p. 182.
42. Ibid., pp. 194–95.
43. See Giffen, p. 26.
44. Charles R. Myer, *Division-Level Communications, 1962–1973* (Washington, D.C.: Government Printing Office, 1982), p. 67.
45. Guy Thomas, "Soviets' Fight-to-Win Doctrine Incorporates Radio Electronic Combat," *Military Electronics/Countermeasures* (December 1982): 40.
46. Spilker, p. 169. This 10 GHz barrier is beginning to be bridged, however. Two of the most recent Air Force space programs, MILSTAR and NAVSTAR, make use of the Extremely High Frequency band (10–100 GHz) for signals passing through the atmosphere.
47. See Dorothy Elizabeth Robling Denning, *Cryptography and Data Security* (Reading, Mass.: Addison-Wesley Publishing Company, 1982), pp. 1–2.
48. Irving M. Gottlieb, "From Smoke Signals to Spread Spectrum, Military Communications Decides Who Wins the War," *Military Electronics/Countermeasures* (December 1982): 44.
49. Ibid., p. 46.
50. Martin P. Ristenbatt and James L. Daws, Jr., "Performance Criteria for Spread Spectrum Communications," *IEEE Transactions in Communication,* COM-25 (August 1977), pp. 760–61.
51. Ibid., p. 756.
52. Ibid., p. 758.

53. Giffen, p. 37.
54. Natural redundancy of the English language, for example, would easily enable the recipient of the garbled message "Plrase sdnd mcney" to correctly interpret it as "Please send money." See Denning, *Cryptography,* pp. 14–16, and Gustavus J. Simmons, "Symmetric and Asymmetric Encryption," *ACM Computing Surveys* (December 1979): 322–23.
55. See, for example, Giffen, p. 37.
56. Bruce A. Smith, "Orbital Survivability of MILSTAR, NAVSTAR Vital to Defense Effort," *Aviation Week & Space Technology* (14 March 1983): 94.
57. Ibid.
58. Covault, "Universe Red," p. 51.

REFERENCES

Aviation Week & Space Technology (7 February 1983): 16.

Bester, Alfred. *The Life and Death of a Satellite.* Boston: Little, Brown, and Company, 1966.

Boffey, Philip M. "Pressures are Increasing For Arms Race in Space." *New York Times,* 18 October 1982. pp. A1, B9.

Broad, William J. "Frying All Our Electrons With a Nuke Over Omaha." *Washington Post,* 5 December 1982. pp. C1–C2.

Canan, James. *War In Space.* New York: Harper & Row, Publishers, 1982.

Covault, Craig. "Space Defense Organization Advances." *Aviation Week & Space Technology* (8 February 1982): 21ff.

———. "Universe Red." *Omni* (August 1980): 49ff.

Friel, Patrick J. "U.S. Space-Based Reconnaissance Communications and Command and Control Systems." Paper presented at the Eleventh Annual Conference on *International Security Dimensions of Space,* sponsored by the International Security Studies Program, the Fletcher School of Law and Diplomacy, Tufts University, 27–29 April 1982.

Giffen, Robert B. *U.S. Space System Survivability: Strategic Alternatives for the 1990s.* National Security Affairs Monograph Series 82-A. Washington, D.C.: National Defense University Press, 1982.

Gjermundsen, Edward I. "Satellite Autonomy—Assuring a Sustaining C3I Capability." *Signal* (February 1983): 23–26.

Gottlieb, Irving M. "From Smoke Signals to Spread Spectrum, Military Communications Decides Who Wins the War." *Military Electronics/Countermeasures* (December 1982): 42–47.

Graham, Daniel O. *High Frontier: A New National Strategy.* Washington: High Frontier, Inc., 1982.

Jones, Merritt E. "DOD's Control Center for Space Shuttle—SOPC." *Signal* (February 1983): 45-54.

Halloran, Richard. "U.S. Plans Big Spending Increase For Military Operations in Space." *New York Times,* 17 October 1982. pp. 1, 60.

Karas, Thomas. *The New High Ground: Systems and Weapons of Space Age War.* New York: Simon and Schuster, 1983.

Lemer, Eric J. "Electromagnetic Pulses: Potential Crippler." *IEEE Spectrum* (May 1981): 41-46.

Macvey, John W. *Space Weapons/Space War.* New York: Stein and Day Publishers, 1977.

Myer, Charles R. *Division-Level Communications, 1962-1973.* Department of the Army. Washington, D.C.: Government Printing Office, 1982. (Document serial 82-22952.)

Ristenbatt, Martin P., and James L. Daws, Jr. "Performance Criteria for Spread Spectrum Communications." *IEEE Transactions in Communication* COM-25 (August 1977): 756-762.

Robling Denning, Dorothy Elizabeth. *Cryptography and Data Security.* Reading, Mass.: Addison-Wesley Publishing Company, 1982.

Simmons, Gustavus J. "Symmetric and Asymmetric Encryption." *ACM Computing Surveys* (December 1979): 305-330.

Smith, Bruce A. "Orbital Survivability of MILSTAR, NAVSTAR Vital to Defense Effort." *Aviation Week & Space Technology* (14 March 1983): 94-97.

Snow, Donald M. "Would Laser Weapons Make the World Safer?" *Christian Science Monitor,* 2 March 1983. pp. 22ff.

Spilker, James J., Jr. *Digital Communications by Satellite.* Englewood Cliffs, N.J.: Prentice Hall, Inc., 1977.

Thomas, Guy. "Soviets' Fight-to-Win Doctrine Incorporates Radio Electronic Combat." *Military Electronics/Countermeasures* (December 1982): 36, 38-41.

Warwick, Graham. "The Military Use of Space." *Flight International* (4 December 1982): 1636-1638.

5 SATELLITES AT SEA Space and Naval Warfare

Louise Hodgden

INTRODUCTION

Sea power, the ability to bring armed force to bear at sea, cannot be determined strictly by the number and kind of warships in a country's order of battle. Measurements of sea power must take into account all of the geographic features, installations, and technologies—weapons and sensors and their support systems—that enable a nation to use force at sea. Any technology that plays an important role in this exercise, sea-based or not, is an instrument of sea power and, when incorporated into the naval force structure, may have far-reaching effects on fleet operating capabilities and manner of fighting.

At present, the most important advances in naval combat capability are not those due to changes in the structure, or outward appearance, of the fleet. Instead, the most salient changes in fleet combat capability have their origins in the modernization of fleet electronics. And of critical importance to these new capabilities are several sophisticated earth satellite systems. This chapter looks at the integration of these space "assets" into fleet operations and ponders the fate of the U.S. Navy should access to these assets, in some future crisis, be lost.

In recognition of this growing importance (and no doubt in part to keep the Air Force from squeezing the Navy out of the business), in

June 1983 Secretary of the Navy John Lehman announced the establishment of the Naval Space Command (NavSpaceCom) to consolidate Navy space and space-related activities. The new command became operational on 1 October 1983. It heralds an important shift in Navy thinking about the role of space in naval warfare, away from the view that space is only peripheral to naval operations and toward the view that space-based systems provide salient advantages in naval communications, navigation, and surveillance.

COMMUNICATIONS

Modern naval warfare is characterized by the requirement for rapid transfer of information. All through the command structure—from the National Command Authority to specified commands, fleets, battle groups, and individual ships—effective force coordination and operational success depend upon swift and reliable communications. Until recently most ship-to-shore and shore-to-ship broadcasts depended upon high frequency (HF) band communications ("short wave"). A ship needing to send a message called the shore station nearest the intended recipient and passed its message on that band. Transmissions from shore stations to the fleet were also sent via high frequency, and ships would listen to all the message traffic, without breaking radio silence, for communications directed to them.

However, HF communications are not always secure or reliable. HF broadcasts bounced off the earth's ionosphere are subject to fading, and there are times in each day when communications are not possible with the desired ship or shore station. An even greater disadvantage is the possibility of compromising a ship's location, and thus its security, with long-range HF transmissions. Direction-finding stations can use HF emissions to locate the position of the transmitting ship. HF broadcasts are also less resistant to jamming than much higher frequency transmissions, for example, in the microwave band.[1]

The importance of HF for ship-to-shore communications has diminished significantly with the advent of communications using just such microwave frequencies. Because their waves propagate in a straight line, do not reflect off the ionosphere, and do not penetrate earth or water to any significant degree, receiver and transmitter must remain within line-of-sight of one another. Satellites in geosynchronous orbits are well suited to these requirements, since they remain constantly within line-of-sight of the same one-third of the earth's surface at all times. With the advent of satellite-to-satellite communications ("crosslinks") in the next decade a task force commander will be able to stay in continuous touch with naval headquarters from any point on the globe without reliance on intervening ground stations and with much less chance of interception or revealing of position. Satellite communications thus have important advantages over traditional HF.

Currently, the FLTSATCOM (Fleet Satellite Communications) system provides multichannel UHF and SHF communications for the Navy. With piggyback Air Force transponders it also supports Air Force bombers and missile launch control centers, airborne command posts, and some parts of the Army nuclear forces. Four satellites in geosynchronous orbit transmit fleet broadcasts to all Navy ships and provide command and control links for exchange of digital data among computers at shore stations and aboard ships (and some aircraft).

The Navy, acting as executive agent for the Department of Defense, awarded a contract in 1978 to Hughes Communications Services to provide a worldwide satellite service to bridge the gap between the end of FLTSATCOM's service life and the initiation of the all-DoD MILSTAR system. This leased satellite system (LEASAT) is due to enter service in 1984 or 1985 and to function for about five years. MILSTAR will replace it, using extremely high frequency (EHF) bands and satellite crosslinks.[2]

In the future, laser communications may offer increased capacity and greater communications security than possible with conventional radio frequency links. Whereas the "footprint" reception area of a satellite broadcasting from geosynchronous orbit might be compressed to a "spot" 300 miles in diameter, the corresponding footprint of a laser system at the same altitude might be only a tenth as much. Such extremely narrow beam widths would provide secure, jam-resistant communications.[3]

Lasercom is also being considered for global communications coverage to submerged submarines. SSBNs never surface and messages are relayed ashore via FLTSATCOM only in grave emergencies since either of these actions could give away their locations to Soviet ELINT satellites. However, because a launch order may be transmitted at any time, the submarine must stay in constant contact with the NCA. To do this currently they deploy a long antenna wire with a buoy at the end that rides below the surface to receive VLF messages from shore stations or TACAMO ("Take Charge and Move Out") communications aircraft. These signals penetrate ocean waters to a depth of several meters. (Normal satellite broadcast wavelengths are too short to penetrate the ocean, so to make use of satellite communications or to monitor FLTSATCOM a submarine must expose an antenna to the air, risking detection.) VLF is not very satisfactory for a number of reasons. VLF transmissions are susceptible to jamming. They are also affected by the electromagnetic pulses generated by nuclear explosions, and their transmitters are highly vulnerable to attack or, in the case of TACAMO, of very limited endurance.[4]

To alleviate these problems, a joint DARPA/Navy Submarine Laser Communications program, also referred to as "Blue-Green Laser," is considering the feasibility of using space-based lasers to communicate with submarines at operational depths. Blue-green lasers are, however, a low priority item because the Navy can reach submerged submarines with ELF transmissions. However, the large antennae for extremely low frequency communications are also quite vulnerable, and ELF data rates are very low—sufficient to transmit launch codes but not for ongoing tactical communication. The decision whether or not to go into full-scale development will be made in 1984, with the early 1990s projected for possible deployment.[5]

It would be desirable, of course, that U.S. naval forces be able to communicate not only among themselves but with allied navies as well. In any future war between NATO and the Warsaw Pact, the side with the less effective communications stands a good chance of losing. Efficient C^3 is especially important to NATO because its command structure is multilayered and places considerable stress on operational flexibility. Its new generation communication satellite (NATO III) is designed to be compatible with the U.S. Defense Satellite Communications System (DSCS) so that U.S. and NATO satellite communications can be integrated and the various national commands can exchange messages. But maritime communications will remain less than adequate at the tactical level. During NATO exercise "Northern Wedding" in 1982, for example, a British frigate equipped with sensitive passive sonar equipment detected a Soviet submarine, but the frigate was not equipped with a satellite terminal. To pass data to the task group commander, an American vessel had to stay in close company with the British frigate and relay its findings.[6] Such disconnects will continue until all NATO navies are equipped with shipboard terminals compatible with FLTSATCOM and DSCS. Whether they are ever so equipped will depend much more on defense economics and alliance politics than on the intrinsic availability of technology.

NAVIGATION

Before satellite navigation systems became operational, sailors used basically the same celestial navigation systems that they had for centuries. Ground-based LORAN (long-range aid to navigation) systems were available but limited in range and area coverage. Moreover, LORAN is affected by certain weather conditions that can alter the timing delays between radio pulses that are critical to position fixing. And solar storms create flares whose electromagnetic effects can disturb the ionosphere and, in turn, radio navigation signals. LORAN signals, for example, reach out to a range of more than 1200 nautical miles at night, with solar effects at a minimum, but only to roughly 600 nautical miles during the day.[7]

The Transit satellite navigation system was originally developed to provide globally available updates to the inertial navigators on board fleet ballistic missile submarines (SSBNs). It has become, over time, a general purpose system for both military and civilian shipping. Military users receiving both Transit frequencies can fix their position with an error of 40 to 400 meters. Positions can be fixed in two dimensions, useful for land and sea-based forces, but not for aircraft or missiles in flight. Moreover, Transit is not continuously available. Time lapses between position updates vary from thirty minutes at high latitudes to one hundred minutes near the equator.[8]

A second-generation network of navigation satellites, the NAVSTAR Global Positioning System (GPS), is expected to reach full operational status by the late 1980s. Three-dimensional position accuracy of about sixteen meters spherical error probable should be obtainable worldwide, with velocity accuracies of 0.1 meters per second. Such capabilities are useful not only for precision navigation. They will substantially affect most naval operations. With NAVSTAR guidance, SLBMs can achieve accuracies necessary for counterforce strikes, allowing greater flexibility in the targeting of the strategic reserve. Conventional weapons (such as the Tomahawk antiship missile) will achieve high accuracies against targets at sea that lie over the horizon, when used in conjunction with surveillance systems able to provide targeting coordinates in real time. Sonobuoys and mines can be dropped and their positions accurately noted for easier retrieval or navigation through the minefields. (Further in the future, long-range, accurately guided aircraft and missiles—and not ships—may provide tactical attack mobility at sea.)

Communications associated with a grouping of forces may reveal the location of those forces. With NAVSTAR, forces could be given the location for rendezvous, disperse if necessary, and concentrate at the desired place with a minimum of radio traffic. Refueling at sea and in the air would become simpler. Naval and air combined operations—with the Rapid Deployment force, for example—would be more easily coordinated, with all forces using the same navigation grid. In addition, NAVSTAR may make it possible for forces to rendezvous under weather conditions, including North Atlantic squalls, where previously the risks of collision or missed position would have been too great. NAVSTAR's coordinate grid will also facilitate search and rescue operations when disasters do occur. Finally, NAVSTAR permits satellite-to-satellite position fixes. To be useful in real-time surveillance, satellites must be able to provide not only data on targets at sea but their precise location as well. If surveillance satellites carried equipment allowing them to determine their own position to a very high degree of accuracy, signals from ships under observation could be immediately correlated with navigation coordinates for precise location fixes and weapons delivery. Seasat A has already carried experimental equipment enabling it to determine its own position to within ten meters by means of signals from NAVSTAR.[9]

OCEAN SURVEILLANCE

The U.S. Navy is developing a space-based surveillance and targeting system to improve its defenses against Soviet ships armed with long-range antiship cruise missiles. Location accuracy and data transfer requirements have become increasingly stringent with the development of such missiles with extended engagement ranges. Over-the-horizon surveillance and targeting is also needed for the Navy's own Tomahawk antiship cruise missile. Operational tests of the Tomahawk have already demonstrated its ability to home in on targets within line of sight, but the Navy still needs over-the-horizon reconnaissance for longer range target coordinates. Apart from the aircraft currently aboard carriers, the Navy has little real-time ocean surveillance capacity. The Navy does have other intelligence sources, including satellites, that have the potential to provide targeting information once their data is integrated. Two Navy programs, known as "Outlaw Shark" for submarines and "Outlaw Hawk" for surface ships, correlate information from various sensors to provide target location to computers aboard Tomahawk-equipped ships. (Outlaw concepts were successfully employed in tests, but operational systems have proven difficult to deploy.)[10]

Two distinct classes of specialized ocean surveillance satellites—one using active radar and the other listening passively for electronic emissions—may be used to develop targeting data. Radar is one of the best sensors for ocean surveillance because it can penetrate cloud cover pointed toward objects of interest. Radar ocean reconnaissance satellites send out strong pulsed radar signals that locate ships by reflection of these emitted signals. First generation types have been referred to as "blob detectors" because they can find large moving objects at sea but do not have the resolution necessary to distinguish warships from other large contacts.[11]

Passive electronic ocean reconnaissance satellites complement the active radar variety by using passive receivers to intercept the radio and radar emissions of ships and determine their functions. Several satellites can triangulate emissions to locate a ship's position. EORSATs can also be used to measure the characteristics of the signals received, ascertaining information necessary to penetrate fleet defenses and to plan electronic countermeasures.

The first American satellites dedicated to ocean surveillance were launched in April 1976 under the Navy's project White Cloud. A second cluster was launched in December 1977. A carrier vehicle placed three small satellites into close proximity, parallel orbits at an altitude of roughly 1,100 kilometers. One report suggests that the three subsatellites in each cluster fly tethered to their carrier.[12] Data received simultaneously from separate points are combined to plot the speed and course of ships under observation.

Each satellite carries an array of antennas used to receive emissions and pinpoint their source, as well as passive infrared sensors to detect heat from nuclear powered vessels. The satellites in a cluster downlink in real time on different wavelengths to receiving stations, which then transmit the information via communications satellites to the Naval Ocean Surveillance Information Center (NOSIC) where it is correlated with additional information from land- and sea-based sensor systems. Ocean surveillance data are then relayed to other commands ashore and afloat. As an indication of their capability, satellites—presumably ocean reconnaissance (White Cloud) satellites—were used in 1978 to track forty ships attempting to smuggle marijuana from Colombia to the United States. NOSIC then transferred the information to the U.S. Drug Enforcement Agency.[13]

Satellite ocean surveillance is not yet available for real-time targeting. By the time satellite data are correlated with data from other sensors and a ship's initial location is determined, it can move out of range in any of several directions. For the immediate future, over-the-horizon targeting will still require a local sensor. Air Force E-3A AWACS will, under two new Memoranda of Understanding between the Air Force and the Navy, support Combat Horizon Extension for Navy missions by providing additional surveillance capabilities.[14] In FleetEx 83-1, held in the northern Pacific, AWACS complemented Navy E-2C reconnaissance aircraft in extending fleet radar horizons by several hundred miles.[15] In the future, as part of an interservice over-the-horizon program, AWACS might also provide targeting coordinates for Tomahawk antiship cruise missiles as well as long-range detection and anti-air tracking for the fleet.

The availability of AWACS will not lessen Navy interest in developing a real-time surveillance capability from space under its own control. With only a relative handful of AWACS available to the United States and NATO, other missions may have priority over assisting the Navy in sea control and antisurface ship strikes. The importance to the Navy of having real-time over-the-horizon capabilities under its control becomes greater as the number of Soviet ships with long-range antiship missiles, and the number of U.S. ships armed with Tomahawk, increase in the 1980s.

AIRCRAFT EARLY WARNING

Perhaps the most important naval lesson of the Falklands War was not a new lesson but one relearned: that naval forces cannot operate effectively against hostile aircraft without long-range surveillance and early warning from elevated sensors. Effective battle management is to a large extent a function of warning time, and in the Falkland case, airborne sensors would have given adequate warning of Argentine attack. Longer warning times would be critical to the U.S. Navy. The Soviet Backfire bomber might be able to breach fleet radar nets via supersonic dash to its missile launch point, 250 to 325 kilometers from its aircraft carrier target. Current E-2C early warning aircraft can "see" only 400 to 475 kilometers: about ten minutes warning against a low-level attack. Yet current U.S. plans are to concentrate maximum response against incoming bombers *before* the launch of their antiship missiles. Satellite-based sensors that provided early warning of an incoming raid would give more opportunity for defensive countermeasures and placing of fighter-interceptors along the most likely axes of attack.[16]

Early warning of an incoming attack becomes even more important as surface warships gain the ability to engage (with cruise missiles) both fixed and moving targets independently of the aircraft carrier and its air wing. However, to operate without a carrier in medium-to-high air threat areas these surface action groups will require some form of early warning beyond that provided by the LAMPS III (light airborne multipurpose system) helicopter carried by many surface combatants. Their operational radius is even more limited than that of the E-2C, and they may be committed in any case to ASW operations instead of AEW patrols. Therefore, nonorganic surveillance systems are needed for early warning of incoming attacks. AWACS may be useful in this role, but they may be committed in wartime to other higher priority missions. As discussed previously, their potential availability does not lessen Navy interest in having real-time over-the-horizon capabilities under its control and priority assignment. Therein lies one of the primary reasons the Navy is interested in a space-based surveillance system to provide location and targeting data on enemy aircraft. The introduction of longer-range surface-to-air (SAM) missiles, with targeting coordinates and in-flight guidance provided by satellite, will permit ship-launched interception to compete with air interceptors like the F-14/Phoenix combination. The use of long-range SAMs would distribute capabilities across the fleet, thereby adding more variables to complicate the enemy's attack calculations.

The infrared signatures of aircraft, surface, and sea targets have already been determined under HI-CAMP (High Resolution Calibrated Airborne Measurement Program). Information from the HI-

CAMP project is being used to develop the infrared sensors for the experimental Teal Ruby satellite, which will test the feasibility of aircraft detection from space. Originally slated for launch in November 1983, Teal Ruby is now scheduled to be launched into low earth orbit by the space shuttle in 1986. A second generation infrared array sensor will be launched later by the shuttle into synchronous orbit. This satellite would be designed to detect aircraft, track them, and determine their flight path with the result that future oceanic warning and combat zones may extend over 1,500 kilometers.

The Navy is focusing a great deal of attention on what it calls "the outer air battle" zone and on command and control and movement of data to support engagements encompassing such vast ocean areas. Because of this requirement, the Navy is moving to develop an Integrated Tactical Surveillance System (ITSS) that will blend approximately thirty different sensor programs. One of the most important of these is a space-based radar.[17] A synthetic aperture radar (SAR) has already been tested aboard NASA's Seasat A, which scanned the earth with a radar swath one hundred kilometers wide. The system produced pictures with a ground resolution of twenty-five meters, irrespective of cloud cover. Rough images were obtained of ocean waves, icebergs, and ships. A companion imaging radar was carried aboard the space shuttle in late 1981, with equally good results.[18] The Navy awarded design contracts in late 1980 for a satellite program even more ambitious, using active radars to spot Soviet bombers in flight.[19] This new program will take on even greater importance to the Navy as the Soviet maritime bomber force equipped with long-range air-to-surface missiles expands to pose a greater threat to the surface fleet.

INTEGRATING DATA: THE CASE OF ASW

Satellite-linked integration of data from forces at sea with shore-based data processing facilities and commanders is particularly significant in antisubmarine warfare (ASW). A key problem in ASW is the analysis of sonar data and how it is affected by propagation parameters—the speed sound waves travel in the ocean, the extent to which they are being bent as they pass through various thermal layers, the number of bounces between the surface and the floor of the ocean, and so forth. Acoustic propagation is strongly influenced by various physical and chemical properties of sea water, particularly by temperature, salinity, and the current. The Navy, therefore, has devoted considerable effort to gathering data on these phenomena and devising global data collection systems. Seabed surveillance systems help to locate and identify sources of unwanted noise. Satellites monitor weather, solar activity, sea states, and other oceanographic data, and identify and eliminate noncombatant vessels from consideration by tracking their electronic emissions.

Once the propagation parameters are determined, submarine acoustic emissions can be identified by passive seabed sonars and the information relayed in real time to computers for correlation with information from other sources.

Mobile forces using active or passive sensors provide tactical antisubmarine reconnaissance. ASW aircraft use a variety of sensors, including magnetic anomaly detection (or MAD, one source of the requirements for accurate, satellite-derived data on the earth's magnetic field), sonar (from sonobuoys), temperature measurements, and forward-looking infrared (FLIR). Ship-based helicopters use sonobuoys or dunking sonar, and some carry MAD. Attack submarines use hull-mounted or towed array sonars.

Once these mobile sensors make contact and the submarine is localized, attack may begin. In practice, however, the problem may be tougher than portrayed here, since "background" traffic can mask the sounds of hostile submarines. In the future, precision navigation updates from the NAVSTAR Global Positioning System will facilitate coordination of such air and sea searches. Currently, a program dubbed "Classic Wizard" integrates data from all sensors via FLTSATCOM and shore-based computers, which return processed information—again via FLTSATCOM—to the fleet.

SPACE, THE SEA AND THE SOVIET UNION

In contrast to standard U.S. Navy operating procedures, the Soviet Navy only deploys a small portion of its blue-water capable forces outside of home waters at any given time. (The Soviet fleet's home waters are the Barents, Baltic, and Black Seas, and the Seas of Japan and Okhotsk). Instead of emulating American steady-state deployments, the Soviets have adopted an operating policy based on flexible, surge deployments to provide a concentration of forces where and when necessary (as during the October 1973 Middle East War). Yearly naval exercises held by each of the four fleets reflect this policy. Most exercises are still held within 2,400 kilometers of the home fleet areas – in the Norwegian Sea, eastern Mediterranean, northwestern Pacific Ocean, and Philippine Sea—but some have been worldwide in scope. In 1970 and 1975, respectively, the *Okean* and *Vesna* exercises encompassed the central Atlantic, central and western Mediterranean, northwestern Indian Ocean, and central Pacific simultaneously, with central coordination of forces.[20]

A Soviet naval priority, stressed by these exercises, is the wartime interdiction of the West's surface-based strike systems capable of launching attacks against the Soviet Union. At present, this threat comes from tactical aircraft aboard carriers. In the future, it will include Tomahawk cruise missiles based on a variety of surface combatants. The requirement for early interdiction of these forces can only be met fully if they can be located early in the conflict. In three different targeting scenarios—continuous company, meeting engagement, and distant targeting—different combinations of ships, aircraft, and satellites would be employed to fix the positions of Western surface forces.[21]

Continuous company describes the situation in the eastern Mediterranean Sea where the Soviet Mediterranean Squadron is on permanent patrol as a counter to the U.S. Sixth Fleet. Satellite surveillance of this area might not be very useful to Soviet planners because the large amount of ship traffic means more time to analyze data to discern vessel types and missions. By the time targeting data could be relayed to Soviet ships, a potential target could easily move away from the coordinates where its position was last fixed. Thus, up-to-date targeting information on NATO warships in the Mediterranean is instead provided by "tattletale" ships. Data from these units are relayed to shore commands and then to the surface strike forces, cruise missile submarines, and land-based missiles comprising the main striking arm of this operational concept.

A meeting engagement might involve carriers headed to launch strikes from, for example, the southern Norwegian Sea or the western Pacific Ocean. Target location would be provided Soviet forces by a mix of forward ships, aircraft, and satellites. The Western force might be harassed en route by aircraft and submarines, but the main engagement would occur in theaters closer to the Soviet Union. A Western force attempting to reach the northern Norwegian Sea would be subject to air- and submarine-launched cruise missile attack and torpedo attack as it crossed north of the Greenland-Iceland-United Kingdom (GIUK) gap.

Finally, those carriers not posing an immediate threat to the Soviet Union and not covered in the above two scenarios would be targeted for long range attack (the "distant targeting" scenario). Aircraft and satellites would provide coordinates; strikes would be carried out primarily by land-based or submarine-based terminally guided ballistic missiles.

These wartime targeting requirements help explain Soviet efforts to maintain an operational space-based ocean surveillance network. The Soviet Union reached initial operating capability with its satellite ocean surveillance network in 1974. The first operational Soviet ocean surveillance satellite was a RORSAT launched that May, followed by an EORSAT in December 1974. (See Chapter 3 for operational details on these and other Soviet military satellites.)

During the *Vesna* global naval maneuvers, RORSATS were orbited to provide location data to Soviet forces. They reportedly located and tracked simulated Western convoys in the Bay of Biscay and may have helped guide Soviet bombers to target.[22]

However, until 1981 such exercises were generally the only occasions that space-based assets were employed. That year marked the first time that at least one Soviet ocean surveillance satellite was operational throughout the entire year. Then, in the spring of 1982, the Falklands campaign presented an outstanding opportunity for the Soviets to exercise their satellite network against an uncooperative target. But a lapse in capability caused by malfunctions in both types of surveillance satellites prevented their use in the early stages of the war. Instead, the Soviets had to use aerial reconnaissance and Soviet naval vessels to shadow the British task force as it sailed to the South Atlantic. This proved unsatisfactory: *Bear* reconnaissance aircraft operating from Cuba and West Africa could not reach the Falklands; intelligence-gathering ships also operating from West Africa were warned away by the British task force; and Soviet ships kept out of the 200-mile war zone around the Falklands (demonstrating Moscow's reluctance to send its forces into a high-risk situation).[23]

Soviet space-based ocean surveillance was not reconstituted until after the British task force had arrived in the South Atlantic, with the launch of passive and active surveillance satellites in late April and mid-May. A second RORSAT was launched in late May for what apparently were synchronized operations with the first pair. By the time of the Autumn Forge 82 NATO exercises, which included large naval maneuvers, the Soviet ocean surveillance program managed to maintain two satellites of each class operational.[24]

The Falklands campaign demonstrated the difficulties of substituting terrestrial for space-based reconnaissance assets for open ocean surveillance of uncooperative targets. That experience may influence Soviet planners to execute naval missions that rely on space-based systems at the onset of war. Any such attack would, logically, occur in conjunction with an attack against the American fleet's own space-based support systems.

The heightened fleet vulnerability that derives from Soviet use of space leads the U.S. Navy to be interested in space for more than ocean surveillance, navigation, precision targeting, and secure communications. The Navy would like to ensure that its satellites continue to transmit data in usable form and that Soviet satellites do not. In 1979, Seymour L. Zieberg, Deputy Undersecretary of Defense for Research and Engineering (strategic and space systems), told Congress: "The principal motivation of our ASAT program is to put us in a position to negate Soviet satellites that control Soviet weapons systems that could attack our fleet. Our ASAT should be principally motivated by the fact that the Soviets have satellites in their force that can trace, locate, and assist in targeting elements of our military forces."[25]

Under certain conditions, U.S. fleet commanders may want to deny the Soviets intelligence on fleet movements (and opportunities to target U.S. task forces) by attacking Soviet ocean surveillance satellites. However, securing the fleet against a Soviet strike would require neutralizing Soviet satellites before the first salvo was fired in order to deny the Soviet Navy the targeting data it would need to plan a first strike. Wartime requirements thus may give both the United States and the Soviet Union incentives to strike the other's satellites preemptively in a crisis.

The tactical advantages of preemption must be weighed against possible strategic costs. Would an attack against fleet support satellites cross a threshold that cannot be crossed without risking further escalation in space and on earth? What might be the implications of attacks on satellites with roles in both conventional/theater and nuclear/global conflict? If the Soviet Union responded against similar U.S. satellites, would the tactical advantages outweigh even the tactical costs?

Electronic countermeasures (ECM) are another way to neutralize threatening satellites, at least temporarily. The orbital paths of ocean surveillance satellites can be tracked and commanders notified when satellites will pass overhead. Countermeasures can then be employed either actively, by radiation of a jamming signal, or passively, by reflecting the enemy's signal from alternate targets ("corner reflectors") designed to mask the real target's location. ECM has the advantage of not (necessarily) damaging the other side's satellites and thus being less likely to provoke a fight if used in periods of stress. Maintaining task group radio silence is another passive method to spoof electronic listening satellites. An aircraft carrier may keep emissions to a minimum even during flight operations. C^3 information can be handled by E-2C aircraft flying at distances from the carrier. Intercepted radio emissions are only traceable to the aircraft, not the carrier. The enemy has to search for the carrier within an area whose diameter is equal to the combined flight radii of the E-2Cs operating with the battle group.

Active jamming may simply saturate the satellite's receiver with noise. Alternatively, the jammer can radiate a subtly modified replica of the enemy's radar signal so that range and angle errors are put into the receiver or false targets are made to appear. Warships with knowledge about the characteristics of a satellite passing overhead also may be able to deliver enough power to overwhelm the satellite's capability to compensate.

OPERATING WITHOUT SPACE SUPPORT

The degree to which the U.S. fleet can operate without space support depends upon the time frame under consideration. At present, the Soviets can attack satellites in low earth orbit (LEO)—the U.S. Navy's Transit navigation satellites and ocean surveillance satellites—with interceptors launched from Tyuratam taking from one-and-a-half to three hours to reach a targeted satellite. Thus far, the Soviet Union has only demonstrated a capability to operate one interceptor at a time. Any degradation of a satellite constellation would therefore occur over a period of time, not instantaneously. The loss of satellites in LEO would not significantly degrade all U.S. naval operations, *given the current force structure.* An analogy might be drawn to a fisherman's net: Even if one portion of the net has a hole in it, the rest can still be used to catch fish. Nonetheless, under persistent attack, gaps in the system could grow and further complicate operations.

With the loss of Transit satellites, for example, surface ships could continue to sail, but submarines would lose position updates to their inertial navigators. One of the main results of an attack on Transit would be the loss of initial position fixes for SSBNs, which could affect targeting accuracy. Accuracy in the launch of a SLBM depends upon precise knowledge of the submarine's location when the missile is fired.

An attack against fleet ocean surveillance would not, in itself, severely complicate current naval operations. The existing White Cloud constellation is not used in real-time targeting, so local sensors are still required to provide targeting coordinates. As the aircraft carrier is still the centerpiece of most naval actions, airborne early warning would usually be available. EORSATs would have been most useful prior to the outbreak of hostilities when they provided information on enemy ship transmissions needed for planning electronic countermeasures and means of penetrating enemy defense.

Soviet interceptors cannot reach satellites in higher orbits such as NAVSTAR and FLTSATCOM. At present, these satellites are safe from direct, nonnuclear attack. (Of course, all satellites are vulnerable to nuclear bursts.)

In the future, one cannot be as sanguine about the loss of space-based assets. By the end of this decade satellites, in conjunction with other ocean area sensors, will have become essential in the tactical integration of maritime forces over targeting grids extending well over 1,500 kilometers.[26] Instantaneous data collection, transmission, and integration for dispersed forces will depend upon satellites performing the roles discussed in this paper. Surface ships may depend upon satellites to provide real-time targeting coordinates on enemy surface ships, and their missiles may be guided to targets 1,000 kilometers away with NAVSTAR position updating. Satellites may also provide early warning of enemy air attack and guide interceptors to target. Hostile submarines may be more quickly located when satellite relay of data is combined with the almost instantaneous data processing that will be made possible by new generations of computers.

Maintaining these capabilities in wartime will require maintaining access to satellites. Ocean surveillance satellites will be particularly important for surface action groups with limited organic air support, if such groups are to cope with such large targeting grids. Space-based ocean surveillance must therefore be maintained, via quick launch of replacement satellites, or via on-orbit "spares." Spare satellites would ensure satellite availability in wartime whatever the condition of U.S. launch sites.

Constellations in high orbits (MILSTAR, NAVSTAR) will remain less vulnerable to performance degradation. Given a Soviet capacity to handle only one interceptor at a time, at most one satellite from either constellation could be damaged every eight hours or so, and that assumes that the targeted satellite could not be maneuvered out of harm's way. Such a slow rate of attack would permit a number of countermeasures, including threats against launch sites and other Soviet interests. It would also give the U.S. Navy time to bring backup systems into play. A fall-back to HF systems would not completely replace communications satellites, however, particularly in the area of data transmission in real time between far-flung operational forces and U.S.-based data processing centers.

The same general considerations would apply to an attack on NAVSTAR. The loss of one or two satellites would not severely degrade the system, though operations requiring constant navigation updates (as, missile guidance or in-flight rendezvous) could be disrupted. Most forces would have to wait longer for position fixes. The analogy of the fisherman's net is most applicable here. It would take several days to attack a significant fraction of the NAVSTAR constellation. In that time, spare satellites could be launched or maneuvered from parking orbits to reconstitute the system.

THREAT REDUCTION AND SATELLITE SURVIVABILITY

Reducing threats to military satellites by negotiation may serve U.S. security better than an unrestrained ASAT competition. In such a competition, the asymmetries inherent in U.S. and Soviet naval operations and their use of space are critical. The location of the majority of Soviet naval forces in or near Soviet home waters means that Soviet naval operations are generally less dependent on satellites than would be U.S. operations seeking entry into those areas. Alternate ground-, air-, and sea-based systems have been maintained to support the majority of Soviet naval missions. The asymmetry is less pronounced for naval operations in distant waters. There, the Soviet Union will depend more on satellites for communications and intelligence. However, a fundamental asymmetry will persist. The ocean areas of greatest importance to both the United States and the Soviet Union are all on the periphery of the Soviet Union. Soviet forces can operate linked to home bases by short lines of communication, but the United States must operate at the end of long lines of communication and supply. An ASAT exchange that removed or degraded the space-based support systems of both sides would only emphasize this asymmetry and make the balance of forces even more unfavorable to the United States. Relative military advantages accrue to the U.S. Navy with continued access to its satellites. Therefore, the best interests of the Navy may lie not in a race for space weapons but in negotiations designed to ensure the continued viability of its other, more important space systems.

NOTES

1. See Chapter 4, supra.
2. Bruce A. Smith, "Orbital Survivability of Milstar, Navstar Vital to Defense Effort," *Aviation Week and Space Technology* (14 March 1983): 94–97.
3. Lieutenant General Richard C. Henry, "Advanced C^3 Seen Possible With Laser Communication System," *NATO's Fifteen Nations* (February–March 1981): 48.
4. *Challenges for U.S. National Security. Nuclear Strategy Issues of the 1980s: Strategic Vulnerabilities; Command, Control, Communications, and Intelligence; Theater Nuclear Forces.* A Third Report Prepared by the Staff of the Carnegie Panel on U.S. Security and the Future of Arms Control. (Washington, D.C.: Carnegie Endowment for International Peace, 1982), pp. 96–97.
5. Deborah G. Meyer, "Strategic Satellites: Our Eyes in the Sky, Can They Save the World from Armageddon?" *Armed Forces Journal International* (February 1983): 38.
6. Desmond Wettern, "Northern Wedding," *Sea Power* (December 1982): 22.
7. Walter B. Hendrickson, Jr., "Satellites and the Sea," *National Defence* (October 1982): 27.
8. K.D. McDonald, "Navigation Satellite Systems: their characteristics, potential and military applications," in Bhupendra Jasani, ed., *Outer Space—A New Dimension of the Arms Race.* (London: Taylor and Francis Ltd., 1982), pp. 160, 166.

9. "Strategic Anti-submarine Warfare and Its Implications for a Counterforce Soviet Strike," in *World Armament and Disarmament Yearbook, 1979.* (London: Taylor and Francis, 1979), p. 438.
10. Norman Polmar, "Extending the Horizon," *U.S. Naval Institute Proceedings* (December 1982), p. 122.
11. Karl Lautenschlager, "Technology and the Evolution of Naval Warfare," *International Security* 8 (Fall 1983): 47.
12. Jim Schefter, "Spy Photos from Space," *Popular Science* (November 1983): 94.
13. Royal United Services Institute, ed., *Defence Yearbook 1981* (Oxford, England: Brassey's Publishers Limited, 1980), p. 179.
14. L. Edgar Prina, "USAF Assumes Larger Role in Maritime Surveillance, Other Missions," *Sea Power* (June 1983): 13.
15. Ibid., pp. 15–16.
16. James W. Lisanby and Reuven Leopold, "The Fleet of the Future—The Impact of Technology," *Sea Power* (April 1982): 141.
17. Rear Admiral George B. Shick, Jr., interviewed by LuAnne K. Levens and Benjamin F. Schemmer, *Armed Forces Journal International* (February 1983): 48.
18. *World Armaments and Disarmament Yearbook, 1979,* p. 20.
19. James Canan, *War in Space* (New York: Harper and Row, 1982), p. 109.
20. Charles C. Petersen, "Trends in Soviet Naval Operations," in Bradford Dismukes and James McConnell, eds., *Soviet Naval Diplomacy* (New York: Pergamon Press, 1979), p. 49.
21. The three targeting scenarios were developed by Michael MccGwire, "Soviet-American Naval Arms Control," in George H. Quester, ed., *Navies and Arms Control* (New York: Praeger Publishers, 1980), pp. 57–78.
22. David B. Kassing, "Protecting the Fleet," in James L. George, ed., *Problems of Sea Power As We Approach the Twenty-First Century* (Washington, D.C.: American Enterprise Institute, 1978), p. 300.
23. Vojtech Maetny, "The Soviet Union and the Falklands War," *Naval War College Review,* May–June 1983, pp. 48–49.
24. Nicholas L. Johnson, *The Soviet Year in Space, 1982* (Colorado Springs, Colo.: Teledyne-Brown Engineering, 1982), pp. 20–21.
25. Quoted in Lcdr. K.Y. Eichelberger, "A New Duel: Antisatellite Combat in Space," *U.S. Naval War College Review* (May–June 1982): 45. The basic mission for the American ASAT was confirmed in the Reagan administration's space policy statement of 4 July 1982:

    > The primary purposes of a United States ASAT capability are to deter threats to space systems of the United States and its Allies and, within such limits imposed by international law, to deny any adversary the use of space-based systems that provide support to hostile military forces.

"Fact Sheet, National Space Policy," in *Weekly Compilation of Presidential Documents* (12 July 1982): 872–76.

26. Lautenschlager, pp. 44–48.

REFERENCES

Canan, James. *War in Space.* New York: Harper and Row, 1982.

Challenges for U.S. National Security. Nuclear Strategy Issues of the 1980s: Strategic Vulnerabilities; Command, Control, Communications, and Intelligence; Theater Nuclear Forces. A Third Report Prepared by the Staff of the Carnegie Panel on U.S. Security and the Future of Arms Control. Washington, D.C.: Carnegie Endowment for International Peace, 1982.

Eichelberger, Lcdr. K.Y. "A New Duel: Antisatellite Combat in Space." *U.S. Naval War College Review* (May–June 1982).

Hendrickson, Walter B., Jr. "Satellites and the Sea." *National Defence* (October 1982): 26–28, 84.

Henry, Lt. Gen. Richard C. "Advanced C^3 Seen Possible With Laser Communication System." *NATO's Fifteen Nations* (February–March 1981): 46–49.

Johnson, Nicholas L. *The Soviet Year in Space, 1982.* Colorado Springs, Colo.: Teledyne-Brown Engineering, 1982.

Kassing, David B. "Protecting the Fleet." Part four, pp. 293–321. In James L. George, ed., *Problems of Sea Power as we Approach the Twenty-First Century.* Washington, D.C.: American Enterprise Institute for Public Policy Research, 1978.

Lautenschlager, Karl. "Technology and the Evolution of Naval Warfare." *International Security* 8 (Fall 1983): 3–51.

Levens, Luanne K., and Benjamin K. Schemmer. "Interview with Rear Admiral George B. Schick." *Armed Forces Journal International* (February 1983): 46–50.

Lisanby, James L., and Reuven Leopold. "The Fleet of the Future—The Impact of Technology." *Sea Power* (April 1982): 127–145.

Maetny, Vojtech. "The Soviet Union and the Falklands War." *U.S. Naval War College Review* (May–June 1983): 46–55.

MccGwire, Michael. "Soviet-American Naval Arms Control." In George H. Quester, ed., *Navies and Arms Control,* ch. 9, pp. 44–100. New York: Praeger, 1980.

McDonald, K.D. "Navigation Satellite Systems: Their Characteristics, Potential and Military Applications." Part two, paper three, pp. 155–188. In Bhupendra Jasani, ed., *Outer Space—A New Dimension of the Arms Race.* London: Taylor and Francis Ltd., 1982.

Meyer, Deborah G. "Strategic Satellites: Our Eyes in the Sky, Can They Save the World from Armageddon?" *Armed Forces Journal International* (February 1983): 30–38.

Petersen, Charles C. "Trends in Soviet Naval Operations." Chapter Two, pp. 37–87. In N.B. Dismukes and J.M. McConnell, eds., *Soviet Naval Diplomacy*. New York: Pergamon Press, 1979.

Polmar, Norman. "Extending the Horizon." *U.S. Naval Institute Proceedings* (December 1982): 121–124.

Prina, L. Edgar. "USAF Assumes Larger Role in Maritime Surveillance, Other Missions." *Sea Power* (June 1983): 13–17.

Royal United Services Institute. *Defence Yearbook 1981*. Oxford, England: Brassey's Publishers Limited, 1980.

Schefter, Jim. "Spy Photos from Space." *Popular Science* (November 1983): 94–97, 186.

Smith, Bruce A. "Orbital Survivability of MILSTAR, NAVSTAR Vital to Defense Effort." *Aviation Week and Space Technology* (14 March 1983): 94–97.

Stockholm International Peace Research Institute. *World Armaments and Disarmament Yearbook, 1979*. London: Taylor and Francis, Ltd., 1979.

Wettern, Desmond. "Northern Wedding." *Sea Power* (December 1982): 21–28.

The White House. "Fact Sheet, National Space Policy." *Weekly Compilation of Presidential Documents* 18 (4 July 1982): 872–76.

6 SPACE-BASED WEAPONS

Dean A. Wilkening

For the past twenty-five years the military uses of outer space have been confined almost exclusively to information gathering and transmission. The global coverage provided by satellites meant that missions previously conducted on earth could be carried out more efficiently and comprehensively in space. Thus, outer space has become a haven for military communication, navigation, reconnaissance, surveillance, early warning, and meteorological satellites. To date, very few weapons have orbited the earth. In the decades to come this may change. This chapter looks at the technologies and missions for space-based weapons, focusing on their technical feasibility, military utility, and strategic implications. Space-based laser (SBL) and space-based particle beam weapons for ballistic missile defense (BMD) will be discussed in some detail. Antisatellite (ASAT) missions will also be touched upon. Before launching into a discussion of these hypothetical systems, however, we should place space-based weapons in their proper context.

HISTORICAL AND POLITICAL CONTEXT

Orbiting nuclear weapons were among the first space-based weapons proposed in the early 1960s. Since they could be made to pass within

200 kilometers of any point in the United States, their prospective deployment gave rise to considerable fear. However, there are sound reasons why these weapons were never deployed. First, the targets they can strike must lie within a narrow strip defined by the satellite's ground track, the width of which would depend on the amount of fuel available for reentry maneuvers. If the satellite is approaching a target aligned with its ground track, only three minutes elapse from the time the weapon deorbits until it strikes the target. Typically, however, it takes several hours for the weapon to pass over the desired target. By contrast, intercontinental ballistic missiles (ICBMs) strike targets within thirty minutes of launch, and submarine-launched ballistic missiles (SLBMs) can take as little as ten minutes. Furthermore, ballistic missiles tend to be more accurate since the launch point is known with greater precision. Second, command and control of orbital weapons would be less secure than for land- or sea-launched ballistic missiles since the communication links might be susceptible to jamming or spoofing. Third, since satellite orbits are predictable, orbital weapons would be relatively easy targets for ASAT attacks (depending on the weapon's maneuvering capability). Finally, since maintenance would be difficult, orbital bombs might malfunction. At best, the weapon and its nuclear material would be lost; at worst, the weapon might deorbit unannounced and detonate accidentally.[1]

The lack of interest in orbital bombs was codified in the Outer Space Treaty signed by the United States and the Soviet Union, among others, in 1967. This treaty forbids the placement of nuclear weapons, or any other weapon of "mass destruction," in orbit around the earth or stationing such weapons in outer space in any other manner (for a detailed analysis, see the following chapter).

As described in Chapter 3, the Soviet Union has tested a Fractional Orbit Bombardment System (FOBS), which launches a warhead into orbit and then deorbits it before completing one revolution of the earth. Strictly speaking, the FOBS does not violate the Outer Space Treaty since the warhead does not make a complete revolution.

In the area of ASAT weapons, both the United States and the Soviet Union achieved rudimentary nuclear antisatellite capabilities during the 1960s because the antiballistic missile (ABM) interceptors both were developing could also be used against satellites in low orbit. The Soviet Union retains that capability; the United States could, if it wished, regain it. In 1968, the Soviet Union began testing a dedicated nonnuclear ASAT weapon whose desultory development program was assessed in Chapter 3.

Since the dismantling of the one U.S. ABM site at Grand Forks, North Dakota, in 1975, the United States has not had a direct-attack ASAT capability. However, the United States currently is developing a new ASAT weapon called the "miniature homing vehicle" (MHV), to be carried aloft by an F-15 fighter and boosted into space by a small rocket. It destroys its target by collision. By most accounts, this weapon will be considerably more versatile than the Soviet interceptor, though still only capable of intercepts at relatively low altitudes.[2]

Nonnuclear ASAT weapons are not constrained by any arms control agreement. During the Carter Administration ASAT talks were held with little success (as related in Chapter 2). Barring a negotiated ban, effective low-altitude ASAT weapons will probably be developed within the next five years. Geosynchronous intercepts may soon follow, since, for a ground-launched interceptor, this is primarily a question of larger boosters, more extensive space tracking, and more capable terminal guidance. (Nuclear warheads would reduce the terminal guidance requirements by a factor of one hundred.)

One final arms control agreement that has a direct bearing on space-based weapons is the SALT I ABM Treaty signed in 1972. It specifically forbids the development, testing, or deployment of space-based antiballistic missile systems (Article 5).[3] In this chapter we will nonetheless examine space-based systems specifically designed for ballistic missile defense since recent technical advances may make this mission more feasible, necessitating renegotiation or termination of the ABM Treaty if they were to be developed, tested, or deployed.

A particular weapon system may be of ambiguous capability, however. For instance, SBLs designed for ASAT missions would have a limited capability against ballistic missiles. One must at least recognize that political recriminations arising from the deployment of ambiguous systems could threaten to destroy an existing treaty one might otherwise wish to preserve.

PHYSICAL CONTEXT

Outer space is by no means a natural environment for weapons. Several attractive features of space are the global coverage, especially coverage of the Soviet Union, and the fact that satellites are relatively immune to direct attack since few countries possess the technical ability to launch into space. The latter advantage is slowly eroding as more countries acquire an independent launch capability and, in any case, never applied to the Soviet Union. Nor does this immunity extend to satellite ground stations.

On the other hand, several physical properties of outer space make it a particularly poor place to locate weapons. First, space is transparent. Virtually any object in orbit can be located and tracked. The best one can do is make a satellite harder to locate by reducing its infrared emissions and its reflectivity at optical and radar wavelengths (stealth techniques). Attempts to counter such measures, by higher power tracking optics or radar beams, would inevitably follow.

Second, orbital velocities are necessarily very high. As a result, an orbital collision can be catastrophic, making it an effective mechanism for destroying space-based weapons (or any other satellite). An ASAT weapon may be guided to collide directly with its target or fragment itself in the target's path. The would-be target could maneuver out of the way, but limited maneuverability is the third disadvantage of space basing. Since orbital velocities are so great, it takes a good deal of energy to deflect a satellite from its orbit. While rocket engines and fuel can be provided for orbital maneuvers (at a cost in weight), it is not clear that much can be gained since an ASAT weapon can carry enough fuel to home on the target even as it attempts to maneuver.

These three characteristics of outer space imply that space-based weapons are potentially vulnerable to relatively simple ASAT weapons. Passive defense against such attacks would be virtually impossible due to the mass of armor required. Nor will active defense work very well against homing projectiles, as we shall see later on. The vulnerability of space-based weapons can be contrasted with the security of ballistic missile submarines, which enjoy high survivability precisely because the medium in which they move is not transparent. Electromagnetic waves do not penetrate sea water very well, and acoustic searches for submarines are hampered by irregularities in the propagation of sound in sea water and by underwater noise. Moreover, the boats are highly maneuverable.

Finally, outer space is unattractive for basing weapons because U.S. and Soviet weapons would constantly intermingle. There would be no sanctuaries where the opposing sides could marshal their forces out of range of attack. A preemptive attack against a space-based

BMD system could eliminate it in seconds and with it a large fraction of one's strategic defense network (a scenario treated at greater length below).

It is not surprising, then, that for the past twenty-five years space has been used for information gathering and transmission, and not for stationing weapons. The recent interest in space-based weapons stems largely from the application of emerging technologies to missions that are thought to be important, especially ballistic missile defense.

MISSIONS AND TECHNOLOGIES FOR SPACE-BASED WEAPONS

Space-based weapons have been proposed for ballistic missile defense, antisatellite missions, and air defense. Notwithstanding the above four disadvantages to stationing weapons in outer space, the advantage of global coverage makes defensive missions (BMD and air defense) particularly attractive on technical grounds.

Missions

Space-based BMD takes advantage of the leverage gained by attacking multiple-warhead ballistic missiles in their boost phase (that is, while their engines are still burning). Boost-phase intercept offers five advantages: 1) the missile booster presents a large, vulnerable target (the high structural stress in an accelerating rocket makes it relatively easy to destroy); 2) the hot exhaust gases provide a nice infrared signature for tracking; 3) there are fewer targets to contend with since the multiple warheads and warhead decoys have not yet been released; 4) decoys that simulate a booster would be more difficult to construct than reentry vehicle decoys (which can be as simple as balloons); and 5) damage assessment is easy since a damaged booster is likely to veer off course or explode. An unavoidable constraint, however, is that the boost phase lasts only about five minutes.

Space-based BMD systems can try to attack a missile's post-boost vehicle (the warhead "bus") or the individual warheads after they are released, but such mid-course intercepts are more difficult since they lack the advantages listed above. Nevertheless, space-based BMD systems have been proposed for both boost-phase and mid-course intercepts as part of a layered defense. A layered defense refers to a situation wherein the ballistic missiles or individual warheads are attacked by different defense systems at different points along their trajectory. A land-based BMD system is often included for terminal defense. If boost-phase, mid-course, and terminal intercept attempts were each 90 percent effective, the overall effectiveness of the BMD network would be 99.9 percent (0.1 percent "leakage"), assuming each layer operates independently.

Both offensive and defensive ASAT missions have been proposed for space-based weapons. The offensive mission requires the capability to attack an opponent's satellites in both low- and high-altitude orbits. The defensive mission would use space-based ASAT weapons to attack an opponent's ASATs. Space-based ASAT weapons would most likely perform both missions equally well, though offensive missions may require weapons of greater range to attack high-altitude satellites.

Finally, space-based weapons may provide some defense against aircraft. Since most of the technologies proposed for space-based weapons do not work very well if the beams or projectiles must propagate through the atmosphere, only high-altitude aircraft would be vulnerable to space-based attack (less than 10 percent of the atmosphere remains above aircraft flying higher than 50,000 feet). Intercontinental bombers would be particularly vulnerable since they must fly most of their mission at high altitude to have sufficient range to strike their targets (low-altitude flight is less efficient by a factor between two and three).[4] Cruise missiles would not be vulnerable to attack from space because they fly too low in the atmosphere. Since air defense is more difficult for space-based systems and since the Soviet bomber threat to the United States is less severe than the ballistic missile threat, this mission will not be discussed further in this chapter.

Technologies

Technologies for space-based weapons fall into two classes: beam weapons (for example, laser or particle beams) and projectile weapons, either rocket- or electromagnetically-propelled.[5] Beam weapons (sometimes called "directed energy" weapons, which is a misnomer since a projectile is also a form of directed energy) are of interest because they can, theoretically, propagate destructive amounts of energy over thousands of kilometers at or near the speed of light. A ballistic missile traveling 3.5 kilometers per second (km/sec) moves only thirty-five meters in the time it takes a laser beam to reach a target 3,000 kilometers distant (0.01 seconds). "Flight time" is not the most important interval to emphasize, however. To cause the desired damage a nominal beam must dwell on its target for about one second. This "dwell time" is more important because it dictates the length of time the target must be tracked and the rate at which different targets can be engaged.

High velocity projectiles would take around 1,000 seconds to reach a target 3,000 kilometers away. If they can be designed to adjust their trajectory in flight, less precise long-distance tracking is required. Furthermore, they impart all of their destructive kinetic energy immediately upon impact. This simple kill mechanism and their relative technical maturity are advantages of projectile weapons compared to beam weapons. Nevertheless, beam weapons have received more attention.[6]

Lasers. "Laser" is an acronym for Light Amplication by Stimulated Emission of Radiation and denotes a device that generates a powerful, collimated beam of coherent light (one in which all the electromagnetic waves oscillate synchronously). Lasers cause damage by focusing the intense light beam onto a target, causing overheating, melting, or even vaporization of the surface material (reminiscent of the holes one can burn in a piece of paper by focusing the sun's rays with a magnifying lens). There are many types of lasers. They vary according to the lasing medium, the mechanism for generating the laser light, and the laser wavelength; each type has distinct advantages and disadvantages (some of which are discussed below).[7]

The most important parameters of a laser are its average light output power (to be distinguished from peak power, for pulsed lasers), the laser wavelength, and the diameter of the optics used to focus and steer the beam. The typical average output power for weapon applications lies in a range from one to one hundred megawatts (MW). Three wavelength regions are of interest for SBLs: the infrared (1–10 microns), ultraviolet and visible light (0.3–0.7 microns), and the X-ray region (about 0.001 microns). The diameter of the optics required to focus the beam is inversely related to the wavelength used (with the exception of X-ray lasers, since X-rays cannot be focused with lenses or mirrors).

The laser most commonly proposed for space-based applications is the hydrogen fluoride (HF) chemical laser, lasing at 2.7 microns (in the infrared). The energy to drive the laser is derived solely from the chemical reaction of hydrogen and fluorine gas. Consequently, no power supplies are required apart from the stored fuel. TRW is currently working on a two-megawatt prototype that can potentially be scaled to twenty-five megawatts.[8] Effective optics for such a laser would need to be on the order of ten meters in diameter, though current plans call for the development and testing of only a four-meter optical system.[9] Since the HF laser shows the greatest near term potential for weapon applications, this system will be used as our analytic model throughout this chapter.

"Excimer" and free-electron devices are two candidates for high-power, visible-wavelength lasers.[10] An excimer laser derives its output from the excited states of an ionized noble gas (for example, xenon or krypton) that reacts to form a lasing medium of xenon fluoride or krypton fluoride. A free-electron laser derives its output from an energetic electron beam as it passes through a magnetic field that varies in magnitude in a periodic fashion along the beam direction. Its wavelength can be varied by varying the energy of the electron beam. Both lasers require external electrical power for their operation, in sufficient quantity that a nuclear reactor would probably be required as the power source. Their chief virtue is their shorter wavelengths, which permit smaller focusing optics (about one meter in diameter). Their chief disadvantage, thus far, is that high power outputs have yet to be obtained.

A current X-ray laser concept involves exposing thin rods of material to the enormous X-ray flash emitted from a nuclear explosion.[11] The explosion X-rays set up the necessary conditions in the rods for coherent (laser) X-rays to be generated. The X-ray laser beams emerge from the ends of the rods only microseconds before the nuclear explosion destroys the whole mechanism. Needless to say, X-ray lasers are single pulse devices. Their advantage lies in the greater lethality of the X-rays and the difficulty of devising countermeasures to them. Their chief disadvantage is their single-shot, self-destructive mode of operation.

Particle Beams. A particle beam weapon is essentially a large particle accelerator, capable of accelerating charged particles to energies on the order of several hundred million electron volts (MeV). If the electrical charge is stripped from the particles as they leave the accelerator, then an energetic neutral particle beam is formed. Neutral beams, as we will see, are essential for space-based weapons.

Particle beam weapons are similar in many respects to laser weapons but they damage targets by hitting them with an intense beam of atomic or subatomic particles instead of an intense beam of light. Their principal advantage lies in the fact that energetic particles penetrate the target's surface, making it harder to devise effective countermeasures. Their principal disadvantage lies in their marginal technical feasibility and their tremendous weight (a key factor in space basing). Electric power sources capable of generating 100 megawatts would be required for these weapons.

Rocket-Propelled Interceptors. These are small missiles capable of boosting either a small explosive warhead to within lethal range of a target or a mass of about five to ten kilograms on a collision course with the target. The MHV is just such a device, though it is launched from an F-15 instead of a space platform. The main attraction of such interceptors is their technical maturity. Further advances in homing sensor technology may bring their accuracy below one meter (circular error probable), so that targets can be destroyed by high kinetic energy collisions.[12]

Electromagnetic Launchers. Electromagnetic launchers, or "rail guns," are devices that accelerate projectiles using electromagnetic forces rather than pressure from hot expanding gases (as in a conventional gun or rocket). The result is higher muzzle velocities (potentially 5–25 km/sec—current devices already achieve 4 km/sec compared to 1.5 km/sec or less for gas driven projectiles). The projectiles destroy their target by collision. The advantage of such systems is the reduced projectile flight time. Major technical problems remain, however, in designing militarily useful devices, particularly with respect to power (peak pulse requirements of 4,000 MW), rapid rate of fire (one round per second is desirable), and projectile homing sensors that can withstand the tremendous accelerations in the gun's "barrel" (as much as 10^5 times the acceleration of gravity at the earth's surface).[13]

Variations on all of the above technologies have been suggested for space-based weapon applications, primarily for BMD, which is generally the most difficult mission to perform. If a system is effective against ballistic missiles, chances are it would have a substantial ASAT capability and perhaps air defense potential as well.

In the following sections the technical feasibility of beam weapons will be considered in greater detail, followed by a discussion of their military utility, especially as defensive weapons, and the strategic implications of their deployment. Technical feasibility is a necessary but not sufficient condition for military utility. Likewise, a militarily useful weapon may not be desirable on strategic grounds. Even though the following treatment focuses on space-based lasers and particle beams, conclusions reached about military utility and strategy are, by and large, applicable to projectile weapons as well.

THE TECHNICAL FEASIBILITY OF BEAM WEAPONS

In this section the basic physical principles behind lasers and particle beams are described so that one can better understand their virtues and limitations as weapons. Approximate values for technical characteristics, which are consistent with the characteristics of proposed systems, are used to describe generic weapons.

Space-Based Lasers

Laser Fundamentals. As mentioned above, a laser is a device that generates a powerful collimated beam of coherent light. Its lethality is determined by the beam's propagation to, and interaction with, a target.[14]

Laser beam propagation depends markedly on the medium through which it passes. If the target is an airplane, the beam propagates part of the way through the upper atmosphere where propagation is complicated by deflection and absorption. Deflection is caused by molecular scattering (the beam's photons ricochet off air molecules), thermal blooming (the beam spreads as it heats the air through which it passes), and atmospheric turbulence (the cause of twinkling stars). Absorption results from the fact that all molecules absorb light at certain characteristic wavelengths. It varies with the precise composition of the atmosphere (which implies that the beam propagation is weather dependent) and with the wavelength of the laser light. At very high light intensities (peak powers greater than 10^7 watts per square centimeter [W/cm^2]) the air becomes ionized. When this happens the beam is completely absorbed by the air. In short, laser weapons are severely constrained when the beam must propagate through the atmosphere.

In the vacuum of outer space laser beams propagate unencumbered by the five mechanisms mentioned above. For a given laser power, the maximum beam intensity that can be focused onto a target is given by the laser power divided by the area of the smallest focused spot that can be formed at the target. The laws of optics define the minimum radius (r) of this focused spot to be,

$$r = (\lambda/D)R, \tag{6.1}$$

where λ is the wavelength of the laser light, D is the diameter of the focusing mirror, and R is the range to the target. This equation defines the "diffraction limit" of the system, a limit that is intrinsic to the wavelike nature of light. All real systems would have radii which are greater than (or equal to) this value.

The maximum beam intensity that can be focused onto a target at range R is given by,

$$I_{max} = P/\pi r^2,$$

where P is the laser output power. Substituting from (6.1) gives,

$$I_{max} = \left(\frac{PD^2}{\pi\lambda^2}\right)\frac{1}{R^2}. \tag{6.2}$$

Note that the variables within the parentheses all pertain to the design of the laser. Thus high beam intensities require high laser powers (P), large focusing optics (D), and short laser wavelengths (λ). Furthermore, the maximum intensity falls off as the square of the distance to the target; that is, targets at twice the range can only be illuminated with one quarter the intensity.

We now turn to the interaction of the focused laser beam with the target. Laser beams incapacitate targets through overheating, damaging vulnerable sensors, melting or vaporizing a hole through their skin, or mechanically puncturing the target from the impulse generated by short laser pulses (which vaporize a thin layer of the target skin at an explosive rate). Impulse loading (the latter mechanism) is the most effective damage mechanism against ballistic missiles.

Only the absorbed energy causes damage; the light that a target reflects causes no harm. The fraction of the laser beam that gets reflected varies depending on the nature of the surface, the laser wavelength, and the intensity of the beam at the target. At low intensity (less than 10^3 W/cm^2) a mirrored surface absorbs only about 1 percent for visible light, whereas metallic surfaces absorb between 3 percent and 10 percent. At high beam intensities (10^7 W/cm^2) the absorption for most surfaces is between 10 percent and 30 percent, due to the creation of a hot ionized gas (plasma) at the target's surface that absorbs the incident laser beam with high efficiency.

To take advantage of this plasma coupling, laser weapons would operate in pulsed mode. The peak intensity of a pulse can greatly exceed the average intensity if the laser pulses are kept short (on the order 0.0001 second). Several pulses may occur each second. For ease in calculating target damage, however, we use average beam intensities throughout this chapter, remembering that the absorption can be as high as 30 percent as a result of the plasma coupling.

Overheating a target results from the absorbed power (as opposed to the absorbed energy). Satellites are particularly vulnerable to overheating since most of their electronics fail above 125 degrees Celsius. As a rule of thumb, satellites overheat if they are exposed to an intensity ten times greater than the intensity of the sun at the earth's surface (0.14 W/cm^2).[15] An intensity of 1.4 W/cm^2 would elevate a satellite's temperature to 400 degrees Celsius after illumination for tens of minutes.

The energy per unit area delivered by a laser beam (known as the "fluence") is equal to the beam intensity (I) multiplied by the dwell time (T). To get the fluence (F) absorbed by the target, one must multiply by the fraction of the laser beam that gets absorbed (a) and by $\sin\theta$, where θ is the angle of incidence. (If the beam is perpendicular to the target, $\theta = 90$ degrees and $\sin\theta = 1.0$; if the beam grazes the target, $\sin\theta < 1$). In other words,

$$F = ITa \sin\theta. \qquad (6.3)$$

Table 6–1 gives the magnitude of the absorbed fluence required to damage different materials. To give a rough mechanical comparison, the muzzle energy of a shell from an M60 tank is approximately 60,000 joules per square centimeter (J/cm^2).

Putting this all together, we can calculate [using (6.2) and (6.3)] the average laser power required, assuming diffraction limited beam

Table 6–1. Fluence Levels for Various Damage Criteria.

Material	*Damage*	*Absorbed Fluence* (J/cm^2)
Eyes	Retinal damage	0.05
Semiconductor	Mechanical damage	10
Human skin	Burns	15
Copper mirror	Pit surface	35
Aluminum (1 mm thick)	Mechanically puncture	1,000
	Vaporize completely	3,500
Thick clouds	Burn through	7,000
Ablative	Mechanically puncture	10,000 to 20,000

Sources: M. Callaham and K. Tsipis, *High Energy Laser Weapons: A Technical Assessment*. Report No. 6 of the Program in Science and Technology for International Security, Department of Physics, Massachusetts Institute of Technology, Cambridge, Mass., November 1980; and M. Callaham, "Laser Weapons," *IEEE Spectrum* (March 1982): 51.

propagation, to cause a specified level of damage out to a range R, for a laser with a given wavelength and mirror diameter. For example, to mechanically puncture the aluminum skin of an ICBM (assumed to be one millimeter thick) at normal incidence, within one second ($F = 1{,}000$ J/cm^2, $T = 1$ sec., $a = .30$, $\theta = 90$ degrees) from a distance of 3,000 kilometers with an HF chemical laser ($\lambda = 2.7$ microns) using a 10-meter-diameter mirror, requires an average laser power output of seventy megawatts. We can now calculate the requirements for an HF chemical laser BMD or ASAT system.

Requirements for BMD. To begin, we need to know the orbital configuration for the SBLs and the number and types of targets against which they are to be effective. If sixty-four SBLs are placed in eight rings (eight satellites per ring) at an altitude of 1,000 kilometers, the maximum range of each SBL must be approximately 3,000 kilometers to ensure complete coverage of the earth. By going to geosynchronous orbit complete coverage can be obtained with two or three SBLs; however, their range must increase to 42,000 kilometers. Since the maximum beam intensity falls off as $1/R^2$, geosynchronous lasers require more than one hundred times the power of SBLs in 1,000-kilometer orbits. Consequently, they are out of the question. Since the boost phase lasts for about 300 seconds and since each SBL should be capable of attacking 300 boosters, a dwell time of one

Table 6–2. Hydrogen Fluoride Laser Power and Fuel Requirements.[a]

Mirror Diameter (m)	*Focused Beam Radius at 3,000 km (m)*	*Target Material*	*Average Power (MW)*	*Fuel Required for 300 Targets at Present [and Theoretical] Efficiencies (metric tons)*	
4	2.0	Aluminum	430	860	[86]
		Ablative	6,400	13,000	[1,300]
10	0.81	Aluminum	70	140	[14]
		Ablative	1,000	2,100	[210]
15	0.54	Aluminum	30	60	[6]
		Ablative	460	920	[92]

[a] The fuel requirements are based on a theoretical specific energy of 1.5 megajoules per kilogram of fuel. The present efficiency is only 10 percent of this value.

Source: M. Callaham, "Laser Weapons," *IEEE Spectrum* (March 1982): 51.

second is allotted for each booster. This system has the potential for defending against the complete Soviet ICBM force (about 1,400 ICBMs) since at any instant five to ten SBLs should be within range of the ICBM fields. Presumably other laser stations would handle SLBM launches.

Table 6–2 gives the HF laser power and fuel requirements for a system capable of attacking 300 ballistic missiles. Three mirror diameters have been used and two levels of target hardness (aluminum skins with $F = 1{,}000$ J/cm^2 and ablative coatings with $F = 15{,}000$ J/cm^2). The beam is assumed to be perpendicular to the target. Systems with larger mirrors require lower power and less fuel, but the mirrors must be pointed with greater accuracy since their focused spot is smaller. For the 10-meter mirror, the beam must track the target with an angular accuracy of 0.3 microradians, which is the angle subtended by a penny when viewed from a distance of seventy kilometers. Also, a mirror can handle only so much power without destroying itself. For 4-meter, 10-meter, and 15-meter mirrors this limit is around 100, 500, and 1,000 megawatts, respectively.

As one can surmise, the 4-meter HF laser is completely inadequate for ballistic missile defense. The 10-meter and 15-meter systems offer some hope. The fuel requirements, while not astronomical, are by no means trivial. A Saturn V booster can lift approximately one hundred metric tons into a 1,000-kilometer circular orbit. Assuming that the mass of the SBL without fuel is small compared to the mass of the fuel, it would require between one and twenty Saturn V-class launches to set up one SBL. Since there must be at least sixty-four SBLs for complete coverage, the launch requirements for a complete network would be between 64 and 1,280 Saturn V-class launches.

Requirements for ASAT. Satellites are more easily damaged than missiles because less intensity is required to cripple their sensors or to cause overheating. On the other hand, the distance to a satellite target may be great, since many military spacecraft are stationed in high-altitude orbits. The effect of these two variables tend to cancel when one is calculating the power requirements for laser ASAT.

The required power can be found by examining three different levels of target damage: *overheating,* which requires a beam intensity of 1.4 W/cm^2 and a dwell time of several minutes; *sensor damage,* which requires 10 J/cm^2 absorbed fluence for semiconductor materials; and *puncturing* the aluminum satellite skin, which requires 1,000 J/cm^2. Again, using (6.2) and (6.3), we can calculate the laser weapon requirements. For example, to overheat an unprotected satellite in geosynchronous orbit requires a five-megawatt HF laser with 10-meter optics. The dwell time would be tens of minutes, unless only a small part of the satellite's mass (for example, the heat-sensitive solar panels) is to be overheated, in which case the dwell time would be on the order of five minutes. Damaging unprotected sensors aboard geosynchronous satellites requires a 40-megawatt HF laser using 10-meter optics, assuming a dwell time of 10 seconds and an absorption coefficient of 10 percent. This same laser could puncture aluminum satellite skins out to a range of 2,300 kilometers. The pointing and tracking requirements for ASAT missions would be the same as for BMD, and the fuel requirements would be similar because of the longer dwell times. However, not more than a few SBLs would be required to attack all satellites of interest.

Technical Assessment. For both BMD and ASAT missions the three following technical requirements must be met: 1) the construction of high-power lasers (for example, 50-megawatt HF chemical lasers); 2) the development of large, diffraction limited mirrors (on the order of ten to fifteen meters in diameter for HF lasers); and 3) the development of pointing and tracking systems that can keep the giant mirrors aimed with an angular precision of 0.3 microradians. The Defense Advanced Research Projects Agency is presently working in each of these areas under the project titles "Alpha" (2-megawatt HF chemical laser), the "Large Optics Demonstration Experiment" (four-meter mirror), and "Talon Gold" (0.2 microradian tracking accuracy).[16]

The technical feasibility of building 50-megawatt HF lasers is probably of least concern, though current technology seems to be more than a factor of ten away from such power levels. The critical problem is to maintain the quality of the laser beam across its ten to fifteen meter diameter.

Were it not for the advent of "adaptive optics," building large diffraction-limited mirrors would be all but impossible. Adaptive optics is a technique whereby a slightly irregular mirror surface is forced into near-perfect curvature by mechanical stress. The largest diffraction-limited mirror for near-infrared wavelengths appears to be the 2.4 meter diameter mirror designed for the Space Telescope. Larger mirrors can, in principle, be built from collections of smaller ones. A difficult problem remains, however. The mirror must remain perfectly shaped despite the tremendous thermal stress it undergoes each time the laser is pulsed. A 50-megawatt laser illuminates a 10-meter mirror with an average intensity of 64 W/cm^2. If the mirror is 99 percent reflective only 0.64 W/cm^2 will be absorbed; however, this is enough to raise its temperature by several hundred degrees centigrade, which would cause severe distortions, especially if there was any temperature differential between different spots on the mirror. Suffice it to say that while building a 10-meter diffraction limited mirror would be difficult, building one to withstand intensities of 64 W/cm^2 would be extremely difficult.

Finally, the technical feasibility of pointing and tracking such large mirrors with 0.3 microradian accuracy seems remote. Current inertial guidance systems can point with an accuracy on the order of 0.01 microradians.[17] The Space Telescope is designed to point toward stars with an accuracy of 0.03 microradian. However, to make a 10-meter mirror (weighing on the order of five metric tons, which must be slewed to engage different targets at the rate of one a second) point with only ten times less accuracy would be quite a feat. Furthermore, to locate the target with this precision requires a 10-meter telescope, if infrared wavelengths are used (to track the infrared emissions from a booster engine, for example), or a two-meter telescope if visible wavelengths are used. Radar or microwave radiation would be completely inadequate to achieve the requisite precision.

Thus, in each of the three key technical areas major improvements are required before the design criteria can be met. The Defense Department estimates that it would take fifteen years before a 10-megawatt, 10-meter SBL would be ready for deployment, and twenty to twenty-five years before a complete constellation of one hundred twenty-five-megawatt, 15-meter SBLs could be deployed.[18] Without a more detailed analysis it is impossible to say whether these time estimates are realistic. At least they are not absurdly short.

Beside the technical feasibility of these components, there is the question of system integration. This involves questions such as whether vibrations in the SBL will degrade the pointing and tracking accuracy, whether the mirror can be kept sufficiently clean so that its reflectivity does not change, whether the SBL can engage successive targets quickly enough, and whether the system can remain quiescent for many years and then suddenly turn on with full power for five to ten minutes. How will the SBL determine the miss distance if the beam fails to hit the target? What about damage assessment? Damaged ICBMs may veer off course or their rocket fuel might explode but satellites may not appear any different after they have been damaged. One could look for the electromagnetic radiation given off when the laser beam heats the target; however, such a signature could be mimicked. Questions such as these must be addressed if the system is to work reliably in outer space.

Space-Based Particle Beams

Particle Beam Fundamentals. As mentioned above, a particle beam weapon is essentially a large accelerator capable of producing charged or neutral beams with 100–1,000 million electron volts (MeV) of energy. Charged particle beams may find some ground-based applications, especially terminal defense for surface ships. However, for space-basing charged particle beams are inappropriate because they bend in the earth's magnetic field, making it difficult to reach distant targets (1,000-MeV proton beams have a 200-kilometer radius of curvature in the earth's magnetic field, 1,000-MeV electrons a 100-kilometer radius of curvature). Furthermore, perturbations in the earth's magnetic field, due either to natural causes (for example, the solar wind) or to a nearby nuclear explosion, would cause unpredictable changes in the beam trajectory. Another disadvantage is that, in the vacuum of outer space, the beam has difficulty propagating due to electrostatic repulsion between the charged particles and to the magnetic field created by the beam current itself.

Neutral particle beams are more useful for space-based applications since they avoid these propagation problems. The most likely candidate is a hydrogen atom (H) beam, which is currently being studied at Los Alamos under the "White Horse" project.[19] Neutral hydrogen beams are formed by accelerating H^- ions in a linear accelerator, then stripping the extra electron off as the beam leaves the accelerator.

For neutral particles the beam diameter at the target is governed by the angular divergence of the beam as it leaves the accelerator. For current high-quality accelerators this angle can be made as low as twenty microradians.[20] But if a beam with an initial diameter of one centimeter is to spread no more than one meter over a distance of 3,000 kilometers, the angular divergence must be no more than 0.3 microradians. As with SBLs, the intensity of a diverging particle beam falls off as the inverse square of the range to the target.

The principle advantage of particle beams over lasers is that the energetic particles penetrate the target rather than interacting with its surface. Thus surface hardening, such as ablative coatings, are ineffective against particle beams. For example, a 200-MeV proton beam (a neutral H beam becomes a proton beam after penetrating a few microns into the target's surface) can penetrate ten centimeters of aluminum. The initial kinetic energy of the beam is transferred to the target material either in the form of heat or in the form of ionized atoms. Thus thermal and radiation damage are the primary damage mechanisms. Depending on the energy deposited per unit volume, various levels of thermal damage may result, chemical explosives may detonate, or the target may begin to melt. Rocket fuel (especially solid fuel) will ignite upon exposure to an intense particle beam. Semiconductors, in particular, are susceptible to thermal and radiation damage. Nuclear warheads can be destroyed by detonating their high-explosive triggers or by radiation damage. Table 6-3 gives values for the absorbed energy per cubic centimeter required to cause the specified damage.

Table 6–3. Absorbed Energy for Target Damage.

Material	*Damage*	*Absorbed Energy (J/cm^2)*
Aluminum	Melt	3,200
Chemical explosive	Detonate	250
Silicon	Melt	7,000
	Transient radiation damage	25 (1×10^6 Rad)
	Permanent radiation damage	1,000 (4×10^7 Rad)

Source: G. Bekefi, B. Feld, J. Parmentola and K. Tsipis, "Particle Beam Weapons–A Technical Assessment," *Nature* 284 (20 March 1980): 219.

We now construct the requirements for a generic space-based neutral hydrogen beam. From Table 6–3 we take 1,000 J/cm^3 to be a reasonable absorbed energy for a wide range of target damage (for either satellites or ballistic missiles). Since a 200-MeV proton beam penetrates approximately ten centimeters of aluminum, a beam one hundred centimeters in diameter at the target must contain a total beam energy of approximately eighty megajoules if 1,000 joules are to be deposited in each cubic centimeter of aluminum (assuming a constant rate of energy deposition). This requires a 0.3 microradian beam divergence if the target is 3,000 kilometers away. To generate these energies requires a pulsed 200-MeV accelerator operating with beam currents equal to 4,000 amperes for 100 microseconds. If this amount of energy can be delivered each second, the average output power would be eighty megawatts.

Since electrical energy is required, the primary energy source for a particle beam weapon could be a nuclear reactor. However, pulsed accelerator operation suggests that a more efficient mechanism might be direct conversion of the energy in chemical explosives (about 4×10^3 joules/gram) to electrical energy.[21] Assuming a 20 percent conversion efficiency from chemical energy to beam energy, approximately one hundred kilograms of chemical explosives would be detonated per second to power the generic particle beam weapon described above.

Requirements for BMD and ASAT. Space-based particle beams are candidates for the same missions as space-based lasers, with the exception of air defense, since neutral particle beams cannot penetrate the upper atmosphere. (The hydrogen atoms become ionized, forming a proton beam that does not propagate in a predictable fashion.)

The generic neutral particle beam described above (80-MW, 200-MeV hydrogen beam with 0.3 microradian beam divergence) would be more effective for boost-phase BMD than space-based lasers since less power is required for equivalent lethality. A constellation of sixty-four space-based particle beams in 1,000-kilometer orbits (eight rings each with eight satellites) would be required for complete coverage. If each particle beam weapon was to attack 300 ballistic missiles within 300 seconds (boost phase), then thirty metric tons of high explosives would be required aboard each weapon for power generation. Alternately, nuclear reactors that can generate one hundred megawatts of electrical power would be required.

Antisatellite missions can also be accomplished more effectively with particle beams. If we assume that all of the beam energy is absorbed by the satellite (that is, the satellite has an effective thickness of ten centimeters or more of aluminum for a 200-MeV neutral hydrogen beam), then the required particle beam intensity to overheat a satellite must be around 1.4 W/cm^2. The required dwell time would be on the order of ten minutes. Overheating geosynchronous satellites requires a 1.6-megawatt hydrogen beam with a beam divergence of 0.3 microradians. Permanent radiation damage to the satellite's electronics can be achieved with beam intensities of 10 W/cm^2 for a dwell time of about fifteen minutes. For geosynchronous satellites this requires an 11-megawatt neutral hydrogen beam with 0.3 microradian beam divergence. For either damage mechanism, approximately ten metric tons of high explosive must be detonated to power the particle beam weapon, for each satellite destroyed. The capability to destroy one hundred geosynchronous satellites would require two or three of these space-based particle beams each with 300 metric tons of high explosive fuel (the space shuttle orbiter Columbia, fully loaded, weights 97.5 metric tons).[22]

Technical Assessment. Questions of technical feasibility for particle beam weapons break into three categories similar to those for space-based lasers: 1) high current, high-energy accelerators (10^4 amperes for 100 microseconds at 100–1,000 MeV); 2) low beam divergence (less than a microradian); and 3) pointing and tracking systems capable of guiding the beam with better than one microradian accuracy. Little information is available for project White Horse; however, recent developments in ion source technology suggest that H^- beam currents of 0.15 amperes are possible.[23] This is over four orders of magnitude less than the requirements for particle beam weapons. Beam energies of 200 MeV can be achieved for H^- accelerators; however, the maximum beam power with current technology is only 0.03 megawatts (versus the required two to eighty megawatts). Furthermore, the smallest beam divergence attainable with current technology is around twenty microradians, a factor of one hundred too large for space-based weapons. The development of high-current, low-divergence particle beams is the key technical obstacle.

As with SBLs, pointing and tracking will be a significant technical problem. However, since particle beams are steered with electromagnetic forces (before the charged beam is neutralized), pointing and tracking may be less difficult than for 10-meter laser mirrors. Nevertheless, 0.3 microradian accuracies require extremely precise electromagnetic fields. Furthermore, the tracking optics and sensors required to locate the target with this precision will be the same as for SBLs. Problems with system integration will also arise.

THE MILITARY UTILITY OF BEAM WEAPONS

The technical feasibility of space-based beam weapons is a necessary but not sufficient condition for their military utility. For a space-based beam weapon to be militarily useful it should be resistant to countermeasures, invulnerable, and cost effective; that is, it should be the least expensive technology for effectively accomplishing the mission and should be less costly to deploy than it is for one's opponent to counter. Since cost estimates for space-based weapons and their countermeasures are difficult to make, given their speculative design, only passing reference will be made to this aspect.

Several relatively simple countermeasures can be used against space-based lasers. Ballistic missiles could rotate during their boost phase, thus spreading the heat over a larger surface area. High-acceleration boosters could reduce the amount of time available for boost-phase intercept. Hardening the missile's surface with ablative coatings has already been mentioned. Table 6–2 shows the jump in laser power required to compensate for ablative coatings (15,000

J/cm^2). Current estimates place the maximum achievable hardness of U.S. or Soviet boosters at 75,000 J/cm^2.[24] New mechanisms for interfering with the laser beam/target interaction (for example, by covering the missile with a film of fluid) might raise the lethal fluence level even further.[25] In any event, one should expect an active competition between laser beams and countermeasures that raise the threshold for damage, similar to the competition between armor-piercing munitions and tank armor. While adding ablative coatings to existing missiles penalizes the missile's throw-weight (for a fixed missile range), the penalty may be acceptable. For a missile the size of the SS-18, doubling the structural mass (total launch mass minus the propellant and payload masses) by the addition of ablative coatings would reduce the throw-weight by 60 percent; that is, from 7,600 kilograms to approximately 3,000 kilograms. For a missile the size of the MX, doubling its lighter structural mass would reduce the throw-weight by 40 percent; that is, from 3,400 kilograms to approximately 2,000 kilograms.[26]

Regardless of the exact mechanism, if missile hardness were doubled (or thought to have been doubled), laser power would have to be doubled to meet the challenge. Otherwise, the lethal range of the SBLs would be reduced by the square root of two, which implies that twice as many SBLs would be needed for complete coverage.

Countermeasures for geosynchronous satellites may be easier to develop due to the relatively low laser intensities that could reach them (less than 10 W/cm^2). A simple mirrored shield could be deployed during attack to reflect the beam and prevent overheating or sensor damage. The shield would have to be destroyed before the laser could illuminate the satellite, a task that requires beam intensities on the order of several hundred watts per square centimeter. Sensors also may be "hardened" by placing filters in front of them which selectively reflect the laser light (though sensors at the focal plane of large telescopes would be very hard to protect since they collect so much light).

Such countermeasures would be largely ineffective against particle beam weapons. Rotating ballistic missiles may help somewhat, but ablative coatings and mirrored surfaces would do little to impede high energy particles. To defend against a 200-MeV neutral hydrogen beam requires a thick shield, weighing 0.3 metric tons per square meter of target surface area—a prohibitive mass for ballistic missiles and most satellites. Sensors and electronics can be radiation hardened to some extent, but sufficient protection is unlikely.

Decoys remain the one viable countermeasure to both laser and particle beam weapons. If space-based beam weapons track the infrared or visible radiation emitted from rocket engines, then infrared or visible flares may be deployed as decoys. In principle, dummy missiles could be launched, though this would be more costly. A similar tactic would be to proliferate the number of missile launchers to saturate a boost-phase defense. A shift toward small single-warhead missiles, as opposed to large MIRVed ICBMs, would be one way to do so.

The vulnerability of space-based weapons is probably their biggest drawback. Space-based weapons are vulnerable to jamming and, as noted before, to physical attack. As with any satellite the communication links can be jammed, though such interference can probably be minimized. A more pernicious form of jamming might involve blinding the sensitive infrared or optical telescopes used by the pointing and tracking system. To give a feel for the problem, the amount of infrared and optical light that reaches a space weapon from a rocket's hot exhaust 3,000 kilometers away is on the order of 5×10^{-12} W/cm^2. The tracking optics must be sensitive enough to detect this. However, an opponent's laser may well be able to direct 10 W/cm^2 into the tracking optics should they be pointed toward the opponent's laser; that is, an intensity over a trillion times as great. To protect these sensors against blinding may be very difficult indeed.

The vulnerability of space-based weapons to physical attack arises from the high kinetic energy associated with objects in orbit, as mentioned in the introduction. A pellet weighing 0.1 kilograms in a 1,000 kilometer orbit has the kinetic energy equivalent to an M60 tank shell. If a cloud of such pellets were released on a collision course with a space-based beam weapon, the weapon could either attempt to vaporize the pellets or maneuver out of the way. Since vaporizing a reasonable mass of pellets consumes a lot of energy, maneuvering is the best alternative. A change in the weapon's velocity by only ten meters per second, if carried out early enough, avoids a pellet cloud several kilometers in diameter. The fuel for several hundred such maneuvers doubles the weight of a space-based weapon.

However, a sophisticated opponent can devise a more complicated attack. Consider the following scenario. Since space-based weapons will be able to maneuver, some form of homing ASAT weapon is required. But homing sensors, especially optical sensors like those used in the miniature homing vehicle (MHV), are vulnerable to beam weapons. Consequently, the ASAT should be protected by a heavy shroud (10–100 kg), to be jettisoned when it is close enough to begin homing on the beam weapon. This shroud can be designed to withstand laser or particle beam fluences on the order of 500 kJ/cm^2. For example, a 15-centimeters-thick shield of aluminum not only takes 500 kJ/cm^2 to vaporize, but is thick enough to stop 200-MeV protons. Obviously in this situation the beam weapon should wait until the shroud is jettisoned before crippling the ASAT's homing device (with relative little effort), then maneuver to avoid colliding with the now inactive ASAT vehicle.

The problem comes when a large number of these homing ASATs are launched from different directions, timed to arrive simultaneously at the target. More than a hundred might be launched at each space-based weapon, since they are cheap by comparison to their targets. A rough estimate can be obtained by noting that the projected cost for a 5-megawatt HF laser with a four-meter mirror is between $2–5 billion dollars, whereas the projected cost for 112 MHVs is $3.6 billion.[27] An interceptor similar to the MHV costing less than $1 million each has been proposed by the Army for mid-course BMD.[28] In this scenario, if the beam weapon waits until the shrouds are jettisoned, insufficient time remains to damage all incoming ASAT weapons. The beam weapon has no choice but to commence firing before the ASAT weapons are within 500 km. (If a 70-megawatt, 10-meter HF laser were to commence firing at 3,000 km, it would take 100 seconds—and one third of the SBL's fuel—to destroy one shrouded ASAT weapon, assuming the latter is closing at a velocity of 10 km/sec.) Consequently, the beam weapon exhausts its fuel or is destroyed in a collision. To aid its defense, the beam weapon could deploy decoys which attempt to fool these homing ASATs.

The beam weapon could instead shoot down the ASAT boosters. In order to separate space launches for other missions from potential ASAT launches, "legitimate" launches must come from one of several predesignated launch sites. All other boosters would be fired upon, especially if there was a large simultaneous launch. Again the fate of the space-based weapon is exhaustion or destruction since it is saturated by 300 or more ASAT launches.

If the vulnerability to a massive conventional ASAT attack is not enough, consider the situation where an opponent has also deployed space-based weapons. Since these weapons are high priority targets, it is likely that each side will stalk the other's beam weapons. If we use a 70-megawatt HF laser with 10-meter optics as an example, we can calculate the shielding required to protect a space-based beam weapon from attack. If an opponent attacks from a range of 3,000 kilometers, a 70-megawatt/10-meter system operating for 300 seconds can deliver a total of one megajoule per square centimeter on target. To shield against this attack requires an aluminum shield ten centimeters thick, or some similarly thick ablative shield. If this shield were to cover the entire weapon (perhaps thirty meters long and ten meters in diameter) it would weigh several hundred tons. If the attack occurred at a range of 1,000 kilometers, the shield would need to be nearly ten times as thick. To protect against a 200-MeV neutral hydrogen beam requires a shield of at least ten centimeters of aluminum. In any event, space-based weapons would look a lot more like metal or ablative cocoons than the fancy pictures one sees in trade journals. Whether one side could actually defeat the other's shielded space-based weapons by striking first would be determined by the details of the measure/countermeasure competition that ensues. While under attack a space-based weapon is pinned down, since it cannot open its shield to fire back. Suffice it to say that the prospect of such a "bolt from the blue" scenario, with little or no tactical warning, could generate considerable fear and instability (considered in the next section). Furthermore, once one space-based weapon is destroyed the effectiveness of a BMD network degrades quickly, since the opponent can time his missile launch to coincide with the hole in the coverage.

In sum, the military utility of space-based weapons is questionable. For space-based lasers relatively simple countermeasures exist. Particle beam weapons (or X-ray lasers) have a considerable advantage in that countermeasures are difficult to devise. Any space-based weapon, however, would be vulnerable to attack. If conventional ASAT weapons were to be used against them in large numbers, space-based beam weapons either become exhausted in their attempts at self-defense or are destroyed by collision with one of the homing ASAT interceptors. An opponent's beam weapons present even greater difficulty since heavy shields would be needed for protection and since no tactical warning would precede an attack. One is tempted to conclude that the side that deployed space-based weapons first would not let an opponent deploy an equivalent capability.

Finally, a word on cost-effectiveness. Space-based weapons for ballistic missile defense may be cost-effective in the sense that few other technologies provide the leverage obtained with boost-phase intercept. They are probably not cost-effective, however, compared to an opponent's ability to deploy countermeasures or to attack them outright. For ASAT missions, space-based beam weapons may be cost-effective compared to ground- or air-launched ASATs;[29] but, once again, countermeasures and direct attack may give an opponent a cost-effective way to neutralize them.

STRATEGIC IMPLICATIONS

Even though space-based beam weapons may not be technically feasible, nor all that useful militarily, is there nonetheless some overwhelming strategic rationale which makes them worth pursuing? To address this question we examine both antisatellite and ballistic missile defense missions. Note that, since the weapon characteristics required for these two missions are not that different, a given piece of hardware may be able to perform both (this will be the perception in any case).

In an ASAT role, space-based beam weapons could deny the Soviet Union the use of those satellites that enhance the effectiveness of its military forces during conflict. Of equal, if not greater importance, is to ensure the survival of those satellites upon which the United States depends (assuming the United States is equally, if not more, dependent on space assets than the Soviet Union). Certainly the preferable situation would be one in which we deploy an effective ASAT weapon and the Soviet Union does not. Such a situation, however, would not last for long. Either both sides will eventually deploy ASAT weapons or both sides will agree to restrain themselves. Achieving U.S. goals (acquiring the capability to destroy Soviet satellites while maintaining the survivability of our own) when both sides have ASAT systems will depend on the details of the measure/countermeasure competition. The outcome may favor the United States, but it will come at considerable financial cost. Alternately, the survivability of U.S. satellites could be enhanced by constraining the Soviet ASAT threat through arms control. The wisdom of this approach depends on the importance of attacking Soviet satellites, since U.S. ASAT weapons would also be constrained. (See Chapter 9 for a full discussion of these issues.)

Ballistic missile defense is by far the more important mission proposed for space-based weapons. Much has been said about how space-based BMD will usher in a new strategic regime promising protection from all types of ballistic missile attack, replacing the current regime wherein the two superpowers are deterred from nuclear attacks on one another by the threat of retaliation in kind. According to the conventional wisdom, such deterrence rests on the prospect that a retaliatory strike would cause unacceptable damage to the aggressor's homeland. Stable deterrence requires the survival of a fraction of one's nuclear forces sufficient to mount such an attack and the political will to do so. If either side's forces can be destroyed in a preemptive attack, or if either country is not vulnerable to nuclear retaliation, deterrence breaks down. This implies that American and Soviet societies must remain vulnerable to nuclear attack, since otherwise a retaliatory threat would be ineffective. Since we never imagine ourselves striking first against the Soviet Union, the need to remain vulnerable to Soviet retaliation seems more of a curse than a requirement for stability. But deterrence became the cornerstone of our nuclear strategy not because strategic thinkers have a penchant for suicide or revenge, but because effective defense against nuclear attack is very difficult to achieve.

Nevertheless, it is reasonable to reexamine our reliance on offense-dominated deterrence as the cornerstone of nuclear strategy and to see whether active defense could play a vital role in changing that strategy. To make this assessment one must lay down goals for nuclear strategy. The three classic goals of arms control can serve equally well, with minor modifications, as goals for strategy: to keep the probability of nuclear war at a minimum; to reduce the consequences should war break out; and to accomplish the first two goals at minimum cost. (If one were to start with different goals, of course, the conclusions drawn might be different.)

How well can strategic defense serve these goals? The most intuitive conclusion is that strategic defenses would reduce the consequences to the defender should war break out. On the face of it this seems desirable, but the problem is not so simple.

The strategic impact of highly capable, space-based defenses would depend on whether the deployment was one- or two-sided. If A deployed first, the effect would be to reduce the consequences to A if war broke out, assuming B took no countering action. This is unlikely, however, because B's ability to deter A rests on his ability to retaliate, which A's defenses diminish. Indeed, with a perfect defense, A could coerce B with impunity. Thus B will react, by deploying countermeasures to render A's defense ineffective, by building more delivery vehicles with which to saturate the defense, by deploying a defense network of his own, or by acquiring the ability to attack A's defenses directly, in space.

Lest one think that attacks against space-based defenses (perhaps in their embryonic stage) would be unlikely, one should recall the events of the Cuban Missile Crisis. Here, the emplacement of Soviet nuclear missiles in Cuba upset the strategic status quo to such an extent that President Kennedy felt it necessary to risk military confrontation in an effort to secure their removal. One of the options put forward by the Air Force, and seriously discussed by the Executive Committee, was a surgical air strike designed to destroy the missiles before they became operational. A naval "quarantine" ultimately was chosen instead. That an effective, one-sided BMD deployment would constitute a serious alteration of the strategic status quo no one should doubt. Since a blockade in outer space is difficult to maintain, a surgical strike might seem attractive.

So, by deploying countermeasures, more offensive forces, or by destroying the defensive system, B can effectively neutralize A's defense. To the extent that B is successful, A has not reduced the level of damage he would suffer in the event of war. A measure/countermeasure competition would be the least destabilizing outcome, since at worst it is a waste of money. If B chooses to deploy more offensive arms, another spiral is added to the arms race. However, A's deployment of space-based defenses could provoke B to take risks (such as attacks on defenses under construction) that could lead to war. Thus, pursuit of the second goal of nuclear strategy, minimizing damage, may violate the first by significantly increasing the likelihood of nuclear war.

What if B deploys an equivalent, highly effective strategic defense, thus undermining A's own ability to retaliate? Strategic security would then rest on the confidence each side had in the effectiveness of its defenses. Despite the difficulties involved in transitioning to such a regime, would the end result nonetheless be worthwhile?

The strategic balance in a defense-dominated world would be rather precarious. Assume each side possesses 10,000 nuclear warheads and both sides have defenses that are 99 percent effective; then, if nuclear war broke out, at most one hundred warheads would penetrate each defense. But if, through deployment of countermeasures, A could reduce the effectiveness of B's defenses to 95 percent, then

500 of A's warheads could penetrate B's defense, giving A a five-to-one advantage in penetrating warheads (a principal measure of the strategic balance under defense dominant conditions). Thus small changes (real or perceived) in defense effectiveness could give rise to large imbalances (real or perceived) in the number of warheads that either side would be able to detonate on the other's homeland. In a defense-dominated balance, the tendency to overestimate an opponent's capability and to underestimate one's own ("defense-conservative planning") would give rise to perennial fears about the strategic balance, making such a world a defense-conservative planner's nightmare. On the positive side, the lack of confidence in one's ability to penetrate the opponent's defense would make decisionmakers reluctant to attack.

In short, the balance of power would be anything but quiescent in a defense-dominated world. Defense technologies would be pushed to the limit at substantial cost, since defensive weaponry is not cheap. The fear of a disarming first strike that dominates current strategic thinking would be replaced by fears of a preemptive strike against defenses or a technological breakthrough which could suddenly reduce their effectiveness.

The argument is made that defenses increase crisis stability by reducing the prospect of a successful first strike.[30] Such an argument fails if the defense is itself vulnerable to preemption. Indeed, vulnerable defenses are particularly destabilizing during a crisis since enormous leverage can be obtained if only a fraction of them can be neutralized or destroyed. A vulnerable defense, in a world of defense dominance, would be as frightening as completely vulnerable offensive forces would be today. (This is not the condition we face today since a large fraction of the U.S. submarine and bomber forces remain invulnerable.)

Ballistic missile defense may appear attractive as a catalyst for reductions in U.S. and Soviet strategic missile forces. It is unlikely to play such a role, however, if only because proliferation of offensive missile launchers is an obvious way to counter an opponent's space-based defenses. Recall that the U.S. decision, in the late 1960s, to produce multiple warhead missiles was in part a product of the desire to saturate Soviet ABM systems.

In closing, it should be noted that space-based BMD could be effective against small attacks or accidental missile launches. Why small powers would not be deterred under present conditions from striking the United States or Soviet Union is unclear. As for accidental launches, there is no doubt that strategic defenses might avert a major catastrophe. This advantage must be weighed against the destabilizing implications outlined above.

CONCLUSION

Despite the current enthusiasm for space-based beam weapons, outer space is not an attractive place to station weapons of any kind. Since space is transparent, space-based weapons can be tracked. High orbital kinetic energies coupled with limited maneuverability imply that space-based weapons would be vulnerable to relatively simple attacks. Finally, the intermingling of opposing space-based forces would give rise to constant fears of attack. Global coverage does convey some advantages for weapon deployments (making boost-phase intercept possible), but it is of greater value for information gathering and transmission, the current military uses of space.

Space-based beam weapons fall short in the areas of technical feasibility, military utility, and strategic stability. They may be technically feasible in twenty to thirty years, if by this we mean that an orbiting weapon could probably direct a sufficiently energetic beam at a ballistic missile several thousand kilometers away and have a reasonable probability of damaging it given present damage thresholds. Daunting engineering problems remain, however. Of the two technologies discussed, space-based laser weapons have greater near-term potential.

The military utility of space-based beam weapons is less obvious, primarily because they would be vulnerable to physical attack. They would surely be high priority targets in a crisis or conflict. The actual effectiveness of such systems would be determined by a fierce measure/countermeasure competition that would probably be more costly for the defense.

Even if one assumes that the technical problems can be solved and that space-based weapons can be made reasonably invulnerable, it is not obvious that one would want them on purely strategic grounds. Self-defense has a visceral appeal, but on closer inspection a defense-dominated regime may be less attractive than the one we have now. The strategic balance would be precarious. Small changes in defense effectiveness could have a large effect on relative offensive capabilities.

The need to hedge against technical breakthroughs would provoke at least increased defense expenditures, and the increased research would promote such breakthroughs. It is not obvious that fear of nuclear war would decrease or that nuclear stockpiles would be significantly decreased, much less eliminated. Finally, there are dangers inherent in the transition from deterrence to defense, even if the end point of highly effective defenses on both sides seems attractive. A one-sided deployment of even a modest BMD system could provoke the other side to take risks that would precipitate war. It behooves us to examine these factors extremely carefully before attempting to alter the foundations of the current nuclear regime. The costs of a mistake are too great.

NOTES

1. U.S. Congress, House, Committee on Science and Technology, Subcommittee on Space Science and Applications, *Space Activities of the United States, Soviet Union and Other Launching Countries/Organizations: 1957–1981.* Congressional Research Service report. 97th Cong., 2d sess., June 1982.
2. William J. Broad, "Weapon Against Satellites Ready for Test," *New York Times,* 23 August 1983.
3. There is some debate on this point due to an ambiguously worded agreed statement (D) associated with the ABM Treaty. Keith Payne and Rebecca Strode explain clearly why space-based lasers would violate the treaty. See "Space-Based Laser BMD: Strategic Policy and the ABM Treaty," *International Security Review* 3:2 (Fall 1982): 269.
4. D. Dommasch, S. Sherby, and T. Connelly, *Airplane Aerodynamics,* Fourth Edition (New York: Pitman Publishing Co., 1967), Ch. 10.
5. Clarence Robinson, Jr., "Panel Urges Defense Technology Advances," *Aviation Week and Space Technology* (hereafter *AWST*) (17 Oct. 1983): 16; Idem, "Study Urges Exploiting of Technologies" (24 Oct. 1983): 50; Idem, "Panel Urges Boost-phase Intercepts" (5 Dec. 1983): 50.
6. For a technical survey of these weapons see *AWST* (28 July 1980): 32; (4 Aug. 1980): 44; and (18 July 1983): 18.
7. K. Tsipis, "Laser Weapons," *Scientific American* (December 1981): 51.
8. *AWST* (28 July 1980): 57; and Clarence Robinson, Jr., "Shuttle May Aid in Space Weapons Test," *AWST* (31 Oct. 1983): 74.
9. Ibid., p. 76.
10. *Physics Today* 36:8 (August 1983): 17.

11. Clarence Robinson, Jr., "Advance Made on High Energy Laser," *AWST* (23 February 1981): 25.
12. Robinson, "Study Urges Exploiting of Technologies": 55, Idem, "Shuttle May Aid in Space Weapons Test": 78; Idem, "Panel Urges Boost-phase Intercepts": 50.
13. Ibid., p. 62.
14. For a technical discussion of these points see M. Callaham and K. Tsipis, *High Energy Laser Weapons: A Technical Assessment,* Report No. 6, Program in Science and Technology for International Security, Department of Physics, Massachusetts Institute of Technology, Cambridge, Mass., November 1980.
15. R. Garwin, "Are We on the Verge of an Arms Race in Space?" *Bulletin of the Atomic Scientists* 37 (May 1981): 48.
16. Clarence Robinson, Jr., "GAO Pushing Accelerated Laser Program," *AWST* (12 April 1982): 16.
17. Callaham and Tsipis. For Space Telescope pointing and tracking see *Physics Today* 36 (November 1983): 47–49.
18. Clarence Robinson, "Laser Technology Demonstration Proposed," *AWST* (16 February 1981): 16.
19. Idem, "Shuttle May Aid in Space Weapons Test": 74.
20. G. Bekefi, B. Feld, J. Parmentola, and K. Tsipis, "Particle Beam Weapons—A Technical Assessment," *Nature* 284 (20 March 1980): 219.
21. Ibid., p. 223.
22. John W. R. Taylor, ed., *Jane's All the World's Aircraft, 1982–83* (London: Jane's Publishing Co., Ltd., 1982), p. 714.
23. Bekefi, et al., p. 222.
24. *AWST* (16 January 1984): 13.
25. Tsipis, "Laser Weapons," p. 57.
26. The SS-18 payload comes from *The Military Balance, 1982–83* (London: International Institute for Strategic Studies, 1982); the MX payload comes from *AWST* (22 March 1982): 18; and estimates of the structural mass come from Hill and Peterson, *Mechanics and Thermodynamics of Propulsion* (Reading, Mass.: Addison-Wesley, 1965), p. 329.
27. U.S. Senate, Congressional Record, 13 May 1981, p. S4978. Description of an amendment on space-based laser weapon research to the Department of Defense Authorization Act for fiscal year 1983; and Broad.
28. Robinson, "Panel Urges Boost-phase Intercepts": 52.
29. Idem, "USAF Will Begin Antisatellite Testing," *AWST* (19 December 1983): 20.
30. For the argument that defenses increase crisis stability see Payne and Strode (supra note 3).

REFERENCES

Aviation Week and Space Technology (28 July 1980): 32 and (4 August 1980): 44.

Bekefi, G., B. Feld, J. Parmentola and K. Tsipis. "Particle Beam Weapons—A Technical Assessment." *Nature* 284 (March 20, 1980).

Broad, William J. "Weapon Against Satellites Ready for Test." *New York Times,* 23 August 1983.

Callaham, M. "Laser Weapons." *IEEE Spectrum* (March 1982).

Callaham, M., and K. Tsipis. *High Energy Laser Weapons: A Technical Assessment.* Report No. 6, Program in Science and Technology for International Security, Department of Physics, Massachusetts Institute of Technology, Cambridge, Mass., November 1980.

Garwin, R. "Are We on the Verge of an Arms Race in Space?" *Bulletin of the Atomic Scientists* 37 (May 1981).

Payne, Keith, and Rebecca Strode. "Space-Based Laser BMD: Strategic Policy and the ABM Treaty." *International Security Review* 3 (Fall 1982).

"Pentagon Studying Laser Battle Stations in Space." *Aviation Week and Space Technology* (28 July 1980).

Physics Today 36 (August 1983): 17–19.

Robinson, Clarence, Jr. "Laser Technology Demonstration Proposed." *Aviation Week and Space Technology* (16 February 1981).

———. "Advance Made in High Energy Lasers." *AWST* (23 February 1981).

———. "GAO Pushing Accelerated Laser Program." *AWST* (12 April 1982).

———. "Beam Weapon Advances Emerge." *AWST* (18 July 1983).

———. "Panel Urges Defense Technology Advances." *AWST* (17 October 1983).

———. "Study Urges Exploiting of Technologies." *AWST* (24 October 1983).

———. "Shuttle May Aid in Space Weapons Test." *AWST* (31 October 1983).

———. "Panel Urges Boost-phase Intercepts." *AWST* (5 December 1983).

———. "USAF Will Begin Antisatellite Testing." *AWST* (19 December 1983).

Tsipis, K. "Laser Weapons." *Scientific American* (December 1981).

U.S. Congress. House. Committee on Science and Technology. Subcommittee on Space Science and Applications. *Space Activities of the United States, Soviet Union and Other Launching Countries/Organizations: 1957–1981.* Congressional Research Service report. 97th Cong., 2d sess. June 1982.

U.S. Senate. Congressional Record, 13 May 1981, p. S4978.

7 THE DEVELOPMENT OF INTERNATIONAL LAW GOVERNING THE MILITARY USE OF OUTER SPACE

Philip D. O'Neill, Jr.

The potential for conflict accompanied mankind's advance into outer space. In the cold war atmosphere of the 1950s, unbridled East-West competition threatened to transform a vacuum into yet another arena for the clash of arms, ideology, and national self-interest. This spectre helped impel nations to seek a common vision of their future relations in a newly accessible environment.

THE INITIAL PHASE: 1957–63

The effort to establish rules for space activities and competition commenced in the late 1950s. In January 1957 President Eisenhower proposed that outer space be dedicated to "peaceful" uses and "denied to the purposes of war." Shortly thereafter, the Western allies offered nuclear disarmament proposals to regulate or to eliminate nuclear weapon delivery of all objects "sent through" outer space, including missiles carrying warheads. The Soviet Union, however, rejected the on-site inspection provisions of the Western disarmament initiatives, linking progress on them to withdrawal of American bases and aircraft from Europe and elsewhere around the Soviet periphery.[1] Soviet coupling of space demilitarization and terrestrial disarmament was strategically unacceptable to the West, as a basis for limiting the military use of space or for moving toward general disarmament on earth. Consequently this initial political sparring failed to harmonize divergent perceptions of proper limits on the growing military use of space.

With the Soviet Union's October 1957 launch of Sputnik I followed by similar American success the next year, the problem of "bombs in orbit" was added along with missiles to the international agenda of space targets for arms limitation. At the same time it became clear that nations were not protesting the use of outer space over their territory as a violation of their sovereignty. Thus, there emerged in embryonic fashion a customary recognition of the freedom to use outer space along with the evident lack of political consensus over its militarization.[2] Ready acceptance by nations of this "free use" principle—akin to that of freedom of the seas—stood in stark contrast to the restrictive attitudes and legal barriers erected over half a century against such use of air space above sovereign territory.[3] Similarly, the short gestation period for the birth of the "free use" and other fundamental principles of conduct in space reached in the 1960s dramatically diverged from the slow, traditional development of international law based on the customs and practices of centuries—reflecting both the impact and pace of technological advance on the law of nations.

During this formative period the United States contributed to the normative process by publicly stating its objectives in space. Congress declared in the National Aeronautics and Space Act of 1958 that American activities should be "devoted to peaceful purposes for the benefit of all mankind" under the direction of a civilian agency.[4] Congress added the caveat that the national security aspects of America's space activities—particularly weapon systems and military operations—would be directed by the Department of Defense. From the outset, then, America's "peaceful" use of space would not necessarily be nonmilitary in nature. Indeed, Congress explicitly recognized that there would be a defense dimension to American efforts in the celestial realm.

The Soviet Union also publicly recognized the military potential of space at an early juncture. Some months after the passage of the NASA legislation, the Soviet Union proposed in the United Nations General Assembly a ban on the use of outer space for military purposes. This public position led the Soviet Union for several years to equate in principle the "peaceful" use of space with its nonmilitary use.[5] The Soviet Union's public position during that period, of course, bore little relation to its actual military program for utilizing outer space. Consequently, Western analysts tended to view the Soviet position as little more than a bargaining chip in the quest for global military advantage, suggesting that it arose from Soviet concern over its slower progress in applying space technology for military purposes. Still, Soviet political rhetoric stood in stark contrast to the position consistently espoused by the United States, that the peaceful use of space encompassed any nonaggressive use, including military measures.

President Eisenhower refined the United States' "militant" posture in a September 1960 address to the United Nations when he proposed a ban on orbiting or stationing weapons of mass destruction in outer space, subject to appropriate verification.[6] The American proposal also separated space from terrestrial disarmament—a segregated environmental approach somewhat analogous to the banning of arms, weapon testing, and military bases from Antarctica in a 1959 treaty to which both the United States and the Soviet Union are parties.[7] The new American proposal did not go as far as the Antarctic treaty, since it contemplated only selective weapon limits and none on the use of space by strategic missiles. Nevertheless, the proposed ban failed to gain immediate acceptance.

By the late 1950s and early 1960s the exploration and use of outer space increasingly became a focus of international concern. During this period, exchanges about Soviet and American military/political space policy and practice found their way into multilateral forums, such as the United Nations, in which non-space powers participated. At the United Nations, the Committee for the Peaceful Uses of Outer Space (COPUOS) became the crucible in which space law was forged through a consensus method.[8] Two years of work in COPUOS preceded the UN's first step toward orderly space exploration. In December 1961, the UN General Assembly unanimously approved a resolution entitled "International Cooperation in the Peaceful Uses of Outer Space."[9] The resolution embraced two cardinal political principles to guide space conduct: (1) international law, including the United Nations Charter, applies to outer space and celestial bodies; and (2) outer space and celestial bodies are free for exploration and use by all states under international law and are not subject to national appropriation.

In the preamble to this UN resolution, nations recognized for the first time the "common interest of mankind" in furthering the peaceful use of outer space and the "urgent need to strengthen international cooperation" there, and declared their belief "that the exploration and use of outer space should be only for the betterment of mankind and to the benefit of states irrespective of the stage of their economic or scientific development." In a less visionary but more practical vein, the resolution's substantive provisions also called upon member states to register with the Secretary General's office all launchings of satellites into orbit. While the space powers compiled with this nonbinding request in reporting launchings, they made no disclosure whatsoever about any military aspects of the satellites' missions.

Although this UN resolution was only a recommendation and did not impose a legal obligation on member states, it nevertheless represented a substantial political limit on the space powers' behavior.[10] The resolution also evidenced basic progress in reaching a consensus among nations over the fundamental political principles of conduct in outer space. In contrast, consideration of military use of space proceeded on a separate negotiating track with less initial success.

In the spring of 1962, both the United States and the Soviet Union included space-related proposals in their general submissions to the Geneva-based UN Committee on Disarmament. The Soviets once again proposed that space be limited to peaceful (nonmilitary) purposes, while the United States resurrected its proposed ban on the placement of weapons of mass destruction in orbit. Canada also proposed a space weapons ban similar to that sought by the United States, but without the crucial on-site inspection provision for disarmament verification. The lack of American support for the Canadian approach soon doomed it. Thereafter, the Soviet Union followed up on these initiatives in the fall of 1962 with a revised proposal that sought a ban on space reconnaissance—thereby seeking to preclude inspection from a jurisdiction outside Soviet sovereign control. America rejected the Soviet Union's proposed limitations and continued to insist upon verification by inspection.[11]

As long as the space powers' proposals remained linked to general disarmament and required on-site inspection, there was no realistic prospect of agreement. However, political incentives for progress on military matters gradually induced a move back toward bilateral exchanges between the Soviet Union and the United States, with a segmented rather than all-encompassing agenda for arms limitation.

Technological advances in verification ultimately created the conditions for a breakthrough in the negotiation logjam. The development of means of inspection from space permitted the United States to drop its on-site inspection requirement in the orbital weapons ban proposal. In short order, the Soviet Union also developed the means to use space reconnaissance for intelligence purposes and eventually dropped opposition to it, without acknowledging the legitimacy of such practices.[12] As a result, the Soviet Union publicly accepted the American-proposed orbital weapons ban at the United Nations in the early fall of 1963, following a clarification that the ban proposal did not seek to regulate the passage of missiles through space.[13] The path to this arms control agreement, while opened by technological advance, was smoothed by the earlier political agreement among nations precluding assertion of sovereignty claims in outer space.

Soviet acceptance of the American proposal followed close on the heels of another political agreement; namely, the Limited Nuclear Test Ban Treaty, which was successfully negotiated late in the summer of 1963.[14] Many nations joined the two superpowers in banning nuclear weapons tests in outer space, in the atmosphere, and under water. This treaty represented the first binding international limitation on the military use of space.

Two months of substantive debate in the United States Senate preceded final American acceptance of the Limited Nuclear Test Ban Treaty.[15] The lengthy ratification process and difficulty in securing approval led the Kennedy administration to adopt a different tactic in securing national commitment to both the proposed ban in space of weapons of mass destruction and comprehensive agreement on political principles governing activities in space.[16] The United States decided to avoid formalizing political agreement in a treaty, choosing instead to rely on a nonbinding UN resolution, with concurrent statements of national intent by America and the Soviet Union.[17] This procedure was advanced in October 1963 despite the Soviet Union's known preference for concluding treaties embodying international agreements and with the knowledge that no legal obligation on either nation or any other would result.[18] Nevertheless, the "conscious parallelism" plan was implemented out of American domestic political concerns and culminated in UN action in the late fall and early winter of 1963.

While the parallel statements of national intent and the UN resolution were nonbinding, they did represent a political commitment to keep weapons of mass destruction out of orbit. Since this approach was not obligatory, there was no need to seek Senate ratification of a treaty or Congressional authorization under the Arms Control and Disarmament Act of 1961,[19] which requires legislative authorization if a president seeks to bind America to limit or to reduce armaments. President Kennedy thus accomplished a de facto moratorium on weapon deployment in space through the exercise of his constitutional powers as chief executive and commander-in-chief. However, neither the United States nor any other nation was legally bound to keep nuclear bombs out of orbit until four years later when a general treaty on outer space with disarmament provisions was concluded and duly ratified.

The basis for a comprehensive space treaty also was laid through the nonbinding approach of a UN resolution. Two years of multilateral negotiations on political issues paralleling the arms control talks were conducted at the United Nations under the auspices of COPUOS from 1961 to 1963. There, nations lead by the space powers gradually moved toward consensus on principles to govern international relations in outer space. Conflicts developed between the Soviet Union and the United States on several fronts following the submission of a Soviet draft declaration of principles, which included prohibitions on the use of outer space to propagate war and for propaganda purposes.[20] Sufficient compromises were ultimately reached by the space powers to permit a consensus on fundamental principles governing international relations in outer space.

The agreed upon principles are enunciated in the General Assembly's December 1963 "Declaration of Legal Principles Governing the Activities of States in the Exploration and Use of Outer Space."[21] The military significance of these principles arose from the political constraints on martial activity that they posed. The first four principles substantially repeated, with some elaboration, those contained in the preamble and substantive provisions of the 1961 UN resolution discussed previously. The fifth principle recognized state responsibility for national space activities, whether performed by governmental or nongovernmental entities. The sixth principle, while mandating international consultation in the event one nation's space activities might cause harmful interference with space activities of another nation, did not provide for a de facto veto as the Soviet Union had originally proposed. The final three principles, which eventually became the subject of specific international agreements discussed later in this chapter, provided for assistance to and return of space vehicles and astronauts (who are regarded as envoys of mankind) as well as for liability for damage from space objects.

Although the peaceful use of space was not included in the substantive provisions of the declaration, the concept was once again recognized in the preamble as being in the interest of mankind. However, the Soviet Union took the position that "[t]he Declaration did not, and indeed could not, touch the use of space for military purposes. . . [for] the U.S.S.R. could not agree to the separation of that problem from other disarmament measures directly related to it, such as the elimination of military bases in foreign territories."[22] As such, the Soviet Union continued to link the demilitarization of space with terrestrial disarmament.

Similarly, the United States adhered to its earlier refusal to accept a ban on the use of space for propagating war. Neither the preamble nor the text of the Declaration contained any such explicit ban. However, a compromise was reached on the balance of the original Soviet-proposed ban by reference in the declaration's preamble to UN Resolution 110(II) of 3 November 1947. That earlier resolution, now made applicable to outer space, condemned propaganda that might provoke a threat, breach of the peace, or some other act of aggression and, further, called upon each state to take steps in this regard "within its constitutional limits." In this way, the United States managed at once to compromise with the Soviet Union in a constitutionally acceptable manner and to limit space as a locus for the broadcast of propaganda that might lead to war, without precluding the use of space in actually making war.

At bottom, the United States regarded the principles set forth in the declaration as a reflection of international law.[23] Together with the Soviet Union, America declared its intention to respect them. As was the case with the earlier "no bombs in orbit" UN resolution, the declaration had no binding effect other than as a limited political commitment and, therefore, did not require Senate or Congressional approval.

Significantly, the UN declaration passed unanimously against the backdrop of substantial Soviet and American military activity in space. Circumstance thus buttressed American claims that peaceful military use of this new environment beyond national appropriation was permissible under international law,[24] which already sanctioned similar peaceful use of international ocean space and air space above it for military maneuvers, weapons testing, and surveillance. In the absence of a binding agreement explicitly precluding any military use of an environment (as existed with respect to Antarctica), the United States could rely on actual state practice in outer space as a basis for the customary development of international law sanctioning its defense posture there. The difficulty presented by such an approach was that the Soviet Union then gave only grudging acknowledgment, if any, to customary practice by nations as a source of binding international law, relying instead upon treaties.[25] There remained, then, an incentive for the United States to move toward conclusion of a binding, comprehensive space treaty with the Soviet Union once American domestic constraints lessened. However, without the necessary political impetus to conclude a treaty, multilateral negotiations in COPUOS languished for several years.[26]

THE OUTER SPACE TREATY

In May 1966, with the United States well along in its program to land men on the moon, President Johnson proposed that the moon and other celestial bodies be demilitarized by treaty.[27] The Soviet Union promptly countered with a proposal by Foreign Minister Gromyko that echoed in large part the American proposal.[28] By mid-June both the United States and Soviet Union submitted to the United Nations draft treaties on space use.[29] Multilateral discussions under the auspices of COPUOS then heated up, particularly in its legal subcommittee, with informal bilateral consultations between the Soviet Union and United States to break negotiation deadlocks.[30] The drafting of the treaty took just six months. In December 1966 the UN General Assembly approved the Outer Space Treaty by unanimous resolution.[31] American ratification of the treaty in 1967 followed Senate debate over verification provisions and executive assurances of the adequacy of our national technical means of inspection, through space surveillance and otherwise.

The Outer Space Treaty transformed a nonbinding, international consensus on political/military conduct into legal obligations and recognized an important new principle: that space was to be the "province of all mankind." More specifically, the Outer Space Treaty tracks much of the same language as the 1963 declaration, including the prohibition of sovereign appropriation of space or any celestial body by any nation, as well as the obligation of states to conduct their activities in space in compliance with international law, generally, and the UN Charter in particular.[32]

The application of the UN Charter to activities in space is potentially significant. Article 2(4) of the Charter obligates nations to refrain from threats or the use of force in international relations against the territorial integrity or political independence of any state. On the other hand, Article 51 explicitly recognizes the right to engage in individual or collective self-defense against armed attack and, by implication, against "imminent" attacks.[33] In practice, then, if one country interferes with the operation of another nation's satellite (over which it retains sovereignty under the Outer Space Treaty), Article 2(4) is breached, and the concomitant right of self-defense under Article 51 is triggered to permit destruction of the source of interference. Thus, in the heavens as on earth, resort to military defensive measures against "aggression" is sanctioned by applicable international law, thereby adding to the complexity of controlling antisatellite weapons (see Chapter 9).[34]

The Outer Space Treaty's arms control measures are set forth in Article 4, which declares that "... [p]arties to the Treaty undertake not to place in orbit around the earth any objects carrying nuclear weapons or any other kinds of weapons of mass destruction, install such weapons on celestial bodies, or station such weapons in outer space in any other manner." This Article also provides, among other things, that "[t]he moon and other celestial bodies shall be used by all States Parties to the Treaty exclusively for peaceful purposes. The establishment of military bases, installations and fortifications, the testing of any type of weapons and the conduct of military maneuvers on celestial bodies shall be forbidden."

These provisions are a landmark in man's attempts to exclude the means of confrontation and war from the celestial realm. The American proposal to outlaw certain weapons from outer space is embraced and celestial bodies, including the moon, are reserved for peaceful purposes and demilitarized in large part. To lend some strength to these measures, a limited right of inspection of objects and installations is provided for in Article 12 with respect to the moon and other celestial bodies, but not outer space.[35]

Still, the treaty's arms control provisions have many shortcomings. The very fact that certain activities are specifically precluded implies to many authorities that all other activities are permitted. While the

Outer Space Treaty excludes nuclear and other weapons of mass destruction from any permanent place in space or on celestial bodies, there is no limitation on such weapons passing through space (e.g., warheads on missiles). Nor is there any limitation on nonnuclear weapons in space other than an acknowledgment that bacterial and chemical weapons that could cause mass destruction are banned.[36]

Moreover, even the limits that are imposed may ultimately prove temporary, since technological breakthroughs could lead a space power to exercise the treaty's one year withdrawal provision set forth in Article 16. Such breakthroughs are possible because the treaty poses no obstacle to testing space weapons except on celestial bodies, for only those places are reserved "exclusively for peaceful purposes."[37] The conspicuous absence of space from the list of domains restricted to such use is not through inadvertance. During the treaty negotiations, India proposed an amendment to the treaty draft that reserved outer space exclusively for peaceful purposes. However, there was insufficient support for this proposal and no consensus was ever reached. As a result, outer space was intentionally omitted by the treaty draftsmen from the celestial domains to be used exclusively for peaceful purposes.[38]

Interpretative difficulties and opportunities for evasion arise even from more explicit treaty provisions. After all, in the absence of definitions, what meaning is to be attached to terms as seemingly straightforward as "in space" or "nuclear weapon" other than that which nations impart to them in practice?[39] In large measure, the problems that arise stem from the simple fact that the treaty and its limits are imprecise statements of principle or goals for behavior. For this reason, nations have sought to give greater definition to behavioral restraints on space activities through a series of bilateral and multilateral treaties.

OTHER SPACE-RELATED AGREEMENTS

In an effort to give concrete expression to general principles set forth in the Outer Space Treaty, nations concluded an agreement in 1968 on the rescue and return of astronauts and objects launched into space. This multilateral agreement spells out the procedures to be followed in situations where "personnel of a spacecraft" or "space objects" come under the jurisdiction or control of a nation other than the launching state. Significantly, no distinction is made between civilian and military use of space personnel or objects. Another noteworthy aspect of this agreement applies to hazardous space objects once they have landed. The treaty obligates the launching state to take effective steps upon notification to eliminate possible danger or harm. This provision does not, however, mandate similar measures while potentially dangerous objects are still in space, thus obviating any treaty obligation for a nation to demonstrate an antisatellite capability against its own space vehicles.[40]

Further progress in limiting the military use of space occurred in the early 1970s when the Soviet Union and United States signed four bilateral accords. Two of these agreements—namely, the agreements on Measures to Reduce the Risk of Nuclear War and the Prevention of Nuclear War, effective in 1971 and 1973 respectively—only incidentally limit the military use of space.[41] Basically these agreements seek to constrain national activity that increases the risk of nuclear war, with its attendant use of space as a medium through which weapons pass. Additionally, the 1971 agreement calls for immediate notification in the event of detection of "interference" with missile warning systems or with related communications facilities (both of which include satellites) if it could create the risk of nuclear war. As such, the 1971 agreement does not preclude interference with space-based facilities or satellites; rather, it seeks only to mandate a consultative step that may help minimize the consequences of interference (whether intentional or accidental) with a nation's space-based sensing and communications devices.

Given the negligible impact of these general measures on specific military uses of space and the lack of any prohibition on interference with space-based monitoring systems, the 1972 SALT I Interim Agreement and Anti-Ballistic Missile Treaty between the United States and Soviet Union took on added importance. Both agreements provide that "each party undertakes not to interfere with the national technical means of verification of the other party. . . ." As such, the legitimacy and legality of using reconnaissance satellites was finally granted public (albeit implicit) recognition by the Soviet Union, at least for the limited purpose of verifying compliance with each treaty. This bilateral recognition did not, of course, obligate other nations to respect the norm. Nor did these 1972 agreements shield from interference or provide legal sanction for satellites performing missions unrelated to treaty verification. With the expiration of the Interim

Agreement by its own terms in 1977 (and the failure to ratify the 1979 SALT II Agreement), the scope of satellite activity legally protected from interference is limited to monitoring prohibitions on antiballistic missile systems. Still, these agreements contributed to the customary development of international space law by indirectly endorsing certain nonaggressive military uses.[42]

Apart from the landmark provisions on verification, the ABM Treaty is highly significant in several other respects from the standpoint of space arms control. Article 5 of the treaty prohibits the development, testing or deployment of antiballistic missile system which are not both fixed and land-based. In addition, the treaty prohibits the deployment of futuristic ABM systems (such as lasers) based on components "capable of substituting" for the ABM components defined in the treaty (e.g., radar, interceptor missiles, and the like). These treaty provisions would certainly be a legal impediment to the space-based missile defenses proposed in the wake of President Reagan's "star wars" speech of March 1983, unless the United States withdraws from the treaty pursuant to Article 15.[43] While the ABM treaty prohibits the development and deployment of futuristic missile defense systems of the sort envisioned by President Reagan, the treaty does not prohibit research and laboratory testing of them. Nor does the treaty prohibit the testing and use of weapons presently under development for antisatellite purposes, so long as there is no upgrade of those weapons to a ballistic missile defense capability.[44] Yet, through the testing of antisatellite weapons considerable information potentially applicable to missile defense can be garnered. At the same time, current attempts to ban totally antisatellite weapons (discussed later in this chapter and in Chapter 9) are hindered, since residual activity allowed under the ABM treaty permits the existence of a ground-based weapon system with the potential for use against satellites in space. Thus, there remain substantial gaps in the existing arms control regime with respect to missile and satellite defense in space.

By the mid-1970s, activity to develop new political limits on the use of space shifted back to multilateral forums. Two additional international agreements developed under UN auspices impose obligations on the space powers. The first agreement is the Convention on International Liability for Damage Caused by Space Objects, which became effective in October 1973. This convention sets forth in considerable detail the substantive rules and procedures governing liability and claims for damage from space objects. Absolute liability, subject to certain exceptions, is imposed upon launching states for terrestrial damage. Thus, this convention puts a price on "free use" of outer space in the event that use results in injury.[45]

In a further multilateral effort to facilitate the orderly administration of space use, the United States and other nations entered into the 1975 Convention on Registration of Objects Launched into Outer Space. Article 4 of the Registration Convention obliges launching states to inform the UN Secretary General about the date and location of a launch, changes in orbit, and the "general" function of the satellite as soon as practicable. By virtue of Article 3, there is "full and open" access to this information. As a result, the mandatory registration provision seems to be more honored in the breach than in compliance, since space powers do not describe any military functions despite the vast number of launches devoted to security or defense purposes.[46]

In nearly a decade of negotiation under UN auspices since adoption of the Registration Convention, two other significant agreements have been reached: the Environmental Modification Convention of 1977 and the Moon Treaty concluded in 1979. The 1977 Convention, which entered into force in 1980, basically prohibits intentional military use (or any other hostile use) of environmental modification techniques to cause destruction, damage, or injury in outer space. The Moon Treaty, on the other hand, generally seeks to close many of the military loopholes left open in the Outer Space Treaty of 1967. For example, Article 3 provides that the moon shall be used exclusively for peaceful purposes; prohibits threats or use of force either on the moon or from it in relation to earth, spacecraft or their personnel, or man-made space objects; and forbids nuclear weapons or other kinds of weapons of mass destruction from placement on the moon, in its orbit, or in any form of trajectory around it.[47]

To date, the Moon Treaty has not been submitted to the U.S. Senate for ratification and remains under study in the Soviet Union. The delay in passage of the treaty results in no small part because of concerns relating to Article 11. That article provides that the Moon and its resources are the "common heritage of mankind" and, further, that the benefits of exploitation are to be shared equitably among nations.[48]

OTHER UN ACTIVITIES

Apart from the 1979 Moon Treaty, the UN's Committee on Peaceful Uses of Outer Space has addressed itself primarily to the issues of direct broadcasting and remote sensing by satellite. Western nations led by the United States have framed the debate about these issues in terms of "freedom of information." The Soviet Bloc and many developing nations have cloaked their opposition in the mantel of state sovereignty, insisting that consent be a precondition to any such satellite use (for further discussion of the Soviet position, see Chapter 8).[49] An international formula for solving these problems remains evasive, however, since practical security concerns restrict the ability of both sides to compromise.

During the late 1970s and early 1980s, COPUOS also sought to develop an international legal norm defining where sovereign air space ends and outer space begins. Concern over this problem received significant impetus when eight equatorial nations claimed sovereignty over the geostationary orbit above their territory in the Bogota Declaration of 3 December 1976, because they feared the loss of exploitable radio frequencies due to the proliferation of communication satellites—particularly by military users.[50] The COPUOS debate over a demarcation line for outer space may ultimately have significant repercussions with respect to the testing and deployment of futuristic space weapons and low-orbit reconnaissance.

Thus far, three discrete positions among nations have surfaced in the debate over delimiting space. Initially, a "functional" approach, keyed to the character of the aeronautical or space activity to be regulated, received support from a number of nonspace powers.[51] The United States and Soviet Union opposed this position arguing that it was premature to seek to delimit outer space and airspace. While the United States and certain other powers still adhere to this view, the Soviet Union in 1979 proposed a "spatial" approach that would fix the boundary line between outer space and air space at 100–110 kilometers above sea level, with a right of "innocent passage" for spacecraft traveling to and from orbit.[52] The United States and other major powers rejected this boundary as arbitrary and continued to oppose all efforts to fix a boundary that technological developments might later render inappropriate. Military considerations seem to motivate this opposition, at least from the American standpoint, since by 1980 some of its close-look satellites were already capable of sustaining orbital perigees approaching the Soviet-proposed boundary. Indeed, if the sovereign airspace boundary exceeded the lowest altitude at which reconnaissance satellites orbit, then such activity would be precluded or subject to defensive attacks similar to those on U-2 flights. As a result, nations were unable to reach a consensus in the early 1980s to establish a legal boundary for outer space or to resolve related sovereignty claims to the geostationary orbit by establishing a legal regime governing its use.[53]

Although political issues largely consumed COPUOS for much of the past decade, the growing military use of outer space did not entirely escape the UN's attention. Contemporaneous with President Carter's initiation of bilateral antisatellite weapon negotiations with the Soviet Union in June 1978 the special session on disarmament of the UN General Assembly appealed to states to undertake further (unspecified) measures through international negotiations to prevent an arms race in outer space.

To this end, in March 1979 at the Geneva-based UN Committee on Disarmament, Italy proposed a protocol to the Outer Space Treaty that sought to reserve outer space for "peaceful purposes only."[54] The draft protocol seeks to mandate that states refrain from participating in measures of a military or otherwise hostile nature; it also seeks to bar "launching into earth orbit. . . objects carrying weapons of mass destruction or any other types of devices designed for offensive purposes. . . as well as the testing of any type of weapon." Both the United States and Soviet Union resisted this multilateral initiative, although it received considerable support from Western and other nations. The American government considered the proposal inopportune in light of its ongoing bilateral negotiations with the Soviet Union and, further, requested clarification of the proposed ban with respect to nonoffensive military space activities.[55]

Similarly, the two space powers resisted a French proposal to the UN Disarmament Committee made in May 1979, which sought to establish a satellite monitoring agency to assist in the verification of arms control agreements. The General Assembly directed that a study be undertaken to ascertain the technical, legal, and financial implications of the proposed agency for future consideration.[56] It remains doubtful, however, that any such agency would have a significant impact on the military use of space by the Soviet Union and United States or impede the proliferation of such use in the near term.

Given the historical linkage between the demilitarization of space and terrestrial disarmament (particularly by the Soviet Union), it is not surprising that the UN's Committee on Disarmament eventually came to be the site of multilateral efforts to head off an arms race in space and to develop international legal norms governing the military use of space. The central importance of that committee's effort was confirmed in 1981 by American opposition in COPUOS to any consideration there of measures to prevent a space arms race. The United States, in contrast to the view expressed by many delegations to COPUOS and echoing the traditional Soviet position, contended successfully that since arms control in outer space was inextricably intertwined with the complex issues of security on earth and arms control in general, consideration of the subject was beyond COPUOS's mandate or expertise.[57]

Thereafter, the UN General Assembly requested unanimously in December 1981 that the Committee on Disarmament consider on a priority basis the question of negotiating an "effective and verifiable agreement aimed at preventing an arms race in outer space...and to prohibit anti-satellite systems...," taking into consideration existing and future proposals.[58] Among the then pending initiatives was a Soviet-proposed draft "Treaty on the Prohibition of the Stationing of Weapons of any kind in Outer Space" presented for UN consideration in August 1981.[59] The Committee on Disarmament considered this proposal in its 1982 session. Little progress was made, however, for the draft protocol contained obvious defects, such as the omission of testing prohibitions, the failure to include ground-launched antisatellite weapons (as opposed to those placed in orbit), and the lack of any requirement to dismantle existing antisatellite weapon systems. Additionally, the proposal that parties to the protocol rely upon national technical means of verification—a standard clause from prior Soviet/American bilateral agreements—had little appeal in a multilateral context since few states had such means at their disposal in the absence of an international verification agency like that suggested by France.[60] The proposed treaty protocol did, however, represent significant movement from prior Soviet positions and protection accorded satellites under bilateral accords with the United States, in that the satellites of all nations that did not carry weapons would be protected from intentional destruction, damage, or other forms of interference (such as disturbing or changing a satellite's normal flight trajectory).

In August 1982, against this backdrop of unsuccessful attempts to resolve political/legal issues or to arrest the growing militarization of space, the United Nations sponsored its second Conference on the Exploration and Peaceful Uses of Outer Space (known as UNISPACE '82). The first such conference, held in 1968, was largely limited to a discussion of scientific and technical issues. The second conference, however, was not so limited, despite efforts by the United States and some other developed nations to avoid the highly politicized atmosphere usually evident in recent ad hoc UN conferences.[61] The Group of 77 issued a declaration on behalf of one hundred and twenty developing nations which threw down the gauntlet to the space powers with respect to the military use of space by opposing the testing, stationing or deployment of any weapons in outer space, calling for adoption of a legal instrument to ban emplacement of weapons subject to verifiable controls, and recommending bilateral negotiations between the United States and the Soviet Union (but without foreclosing consideration of the issues by UN committees).[62] A large majority of nations also sought to ban the testing and deployment of antisatellite weapons and to guarantee the inviolability of all peaceful activities in space.[63]

While nations attending UNISPACE '82 were unable to reach a consensus on most of the substantive proposals relating to military use of space, they did agree in the final conference report that the extension of an arms race into outer space is a matter of grave concern and detrimental to humanity as a whole. The report urged the space powers to assign high priority to preventing an arms race in outer space and to refrain from any action contrary to that aim. The outstanding political/legal issues relating to direct broadcasting, remote sensing, and the delimitation of space remained unresolved.[64]

The recommendations set forth in the final report of UNISPACE '82 contain no magic formulas to resolve long-standing substantive disputes. They do, however, reflect growing political will among non-space powers to press for agreement on international legal rules to end or at least to curtail the development and deployment of outer space weaponry and to resolve political/legal issues on a variety of other fronts. As a part of that process, nations are increasingly isolating the major space powers, particularly the United States, from the publicly accepted views on space use shared by an overwhelming majority of countries. Moreover, as tolerance for the consensus method hitherto employed wears thin, there seems a growing willingness to forsake the benefits of binding agreements based on unanimity for non-binding limitations derived from political pressure in a variety of forms, including UN resolutions.[65]

Thus, the battle to develop international legal norms is shifting out of closed diplomatic circles and into the open court of public opinion. The parameters for the debate over appropriate uses of outer space and the adequacy of the existing international legal regime are already crystallizing.

RECENT AMERICAN AND SOVIET INITIATIVES

In March 1983 President Reagan articulated in his "star wars" speech an American vision of national security which embraces a space-based missile defense system and which seeks to divorce strategic policy from its shotgun marriage with the notion of mutually assured destruction. By mid-October 1983 a senior interagency group recommended to the President that the United States embark on a program to demonstrate as early as possible the practicality of such a system, without impinging in the "near term" on existing treaties, protocols, agreements, and political constraints.[66] President Reagan reportedly agreed to this approach in December 1983 by deciding to emphasize long-term research and development, rather than to embark on a program to deploy defensive space weapons quickly.[67]

The Reagan administration plans for the military use of outer space have not developed in a domestic vacuum. In July 1983 Congress constrained testing of American antisatellite weapons through the Tsongas amendment to the 1983 Defense Department Authorization Act.[68] Senator Tsongas's amendment prohibited tests against objects "in space," except where the President certifies that he is attempting to negotiate a verifiable treaty banning antisatellite weapons and, further, that pending agreement on such a ban, testing against objects in space is necessary to avert clear and irreparable harm to national security.[69] At the same time, both the United States Senate and House of Representatives began moving toward a consensus to adopt a nonbinding, joint resolution calling for immediate negotiations for a ban on the use of force in or from space; weapons of any kind in space; and the testing, production, deployment, or use of antisatellite weapons (including systems based on earth).[70]

Shortly thereafter, in August 1983, a delegation of American senators visited Soviet Chairman Andropov and received a similar proposal for the peaceful use of space. The Soviet proposal, which has been submitted for UN consideration, combines a unilateral Soviet moratorium on launchings of antisatellite weapons with a draft treaty prohibiting the use of force in or from outer space and banning space weapons.[71] Although there are defects in the Soviets' 1983 draft, it nevertheless represents a significant improvement over their 1981 proposal, since it would prohibit testing and ground-based systems, calls for the dismantlement of existing systems, and drops limitations on space object protection tied to the behavior and mission of the satellite.[72] Furthermore, the new Soviet proposal is both responsive to Congressional initiatives and in keeping with recent UN resolutions and the recommendations of UNISPACE '82.

As the Soviet Union moves to join the vast majority of nations and the American Congress in pressing for the conclusion of new agreements limiting the military use of outer space, the Reagan administration's expansive attitude about space as a base or medium to use in defense against terrestrial aggression is undergoing some refinement. However, given the Reagan administration's publicly stated commitment to developing defensive, space-based ballistic missile defense systems, there appears little likelihood that in the absence of domestic legislative or technological constraints, the United States will commit itself in the near future to outlawing the stationing of weapons in space or the use of force in or from that environment. Consequently, unless the Soviet Union is willing to decouple the conceptual linkage of its proposed antisatellite weapon ban with broader prohibitive measures against the use of force in or from space, then prospects for agreement will dim. Given the pace of technological development and verification difficulties inherent in systems such as the American antisatellite weapon now under development, or Soviet space mines, a widely perceived window of opportunity for control of the weaponization of space may soon close. It would seem, then, if there is to be progress or the opportunity for progress in space arms control, the status quo will have to be preserved—perhaps in the form of de facto test bans through "conscious parallelism" similar to that evidenced in the moratorium leading to the 1963 Limited Test Ban Treaty. In addition, the United States and Soviet Union will likely have to return either to a more selective or measured approach to space-weapons control along the lines of the 1972 ABM Treaty (e.g., focusing solely on antisatellite weapons) or to agreements of limited duration (e.g., the SALT I Interim Agreement).

CONCLUSIONS

Regardless of the future outcome of any domestic or international compromise relating to space arms control, the Reagan administration's current actions and positions are compatible with existing international space law and, at least for the time being, with bilateral accords such as the ABM Treaty. Similarly, the Reagan administration's vision of space as a legitimate forum for defensive military measures is, in many ways, perfectly consistent with the public position of its predecessors since the late 1950s. However, the technological advances of recent years coupled with the political will to develop and to deploy space weapons does portend a significant change in America's military posture in space. As a result, the timing of the 1983 space arms control proposal by the Soviet Union reflects, in many ways, not only Soviet sensitivity to our domestic legislative process but also a concern that the Soviet Union has fallen behind the United States in the application of space technology to military purposes—just as President Eisenhower's proposals in 1957 reflected similar American concern over a Soviet missile advantage.

Since the actions of nations are a principal source of international law, American military activity in space will ultimately affect the development of customary norms governing the military use of space. Similarly, an American refusal to abide by nonbinding UN resolutions on space, or to supplement existing treaty obligations to close loopholes, would impede the further expansion of international space law through custom or by agreement. Thus, resolution of the domestic debate in the United States over planned military use of space will have major consequences not only for our national space and defense policy but also for peaceful use of space by all nations.

At bottom, then, the resolution of a quarter century of international debate over political/military principles governing the use of outer space will turn largely on whether the United States ultimately chooses to arm or to preclude armament in space, to allow use of that environment for the projection of force, and to exploit for military advantage the limited, existing international legal regime governing outer space. Because the Soviet Union has apparently offered in a serious manner to forego a space arms race and to preserve space as a weapon-free arena (albeit with attendant verification problems), the choice to pursue or to limit such a race is basically up to the United States. That is the choice with which the next age in space begins; America's response—in deciding whether to confine the means of conflict to earth or to add a new dimension to our society's defense through increased military use of space—will dictate for all practical purposes both the future development of international space law and the nature of "peaceful use" of that environment.

NOTES

1. See, generally, Bernard G. Bechhoffer, *Postwar Negotiations for Arms Control,* (Washington: Brookings Institution, 1961), pp. 395, 473, 475, and 545–46.
2. Many Western international legal scholars include custom among the recognized sources of international law. See, for example, Michael Akehurst, "Custom as a Source of International Law," 1974–75 *British Year Book of International Law* (Oxford: Clarendon Press, 1977), pp. 1–10. With respect to the customary freedom to use outer space, see the UN Ad Hoc Committee Report on the Peaceful Uses of Outer Space. U.N. Doc. A/4141/25 (1959).
3. Each state retains complete and exclusive sovereignty over the airspace above its territories. See, for example, Article I of the 1944 Chicago Convention on International Civil Aviation, 61 Stat. 1180; T.I.A.S. No. 1591; 15 U.N.T.S. 295.
4. 42 U.S.C. §2451.
5. The draft resolution is dated 7 November 1958. See U.N. Doc., G.A. Verbatim Off. Rec., 13 Sess. 1st Comm. A/C.1/L. 219 (1958). See Emilio Jaksetic, "The Peaceful Uses of Outer Space: Soviet Views," *American University Law Review* 28 (1979): 483, 493–98.
6. U.S. Senate, Committee on Aeronautical and Space Sciences, *Statements by Presidents of the United States on International Co-operation in Space—A Chronology: October, 1957–August, 1971,* S.Doc. No. 92-40, 92d Cong., 1st sess. 16 (1971).
7. U.S. Arms Control and Disarmament Agency, *Arms Control and Disarmament Agreements* (Washington, D.C.: U.S. Government Printing Office, 1980), pp. 19–27 (hereinafter "Arms Control Docs."). T.I.A.S. No. 4780; 1 U.S.T. 794; 402 U.N.T.S. 71.
8. See, generally, Michael Bourely, "The Contributions Made by International Organizations to the Formation of Space Law," *Journal of*

Space Law 10 (1982): 139, 140–47 (hereinafter "J.Space L."). See also Eilene M. Galloway, "Consensus Decision-Making by the United Nations Committee on the Peaceful Uses of Outer Space," *J.Space L.* 7 (1979): 3.

9. U.N.G.A. Res. 1721/XVI, Dec. 20, 1961.
10. See, for example, the memorandum by the Office of Legal Affairs in the UN Secretariat on the nonbinding nature of UN recommendations and declarations. U.N. Docs. E/CN. 4/L 610 (2 April 1962).
11. See, generally, Raymond L. Garthoff, "Banning the Bomb in Outer Space," *International Security* 5 (Winter 1980/81): 25, 27–28.
12. See, for example, U.N. Docs. A/AC. 105/C.2/L.6 at 2 (1963). See D.G. Brennan, "Arms and Arms Control in Outer Space," in L.P. Bloomfield, ed., *Outer Space—Prospects for Man and Society* (Englewood Cliffs, N.J.: Prentice-Hall, 1962), Ch. 6, pp. 161–62.
13. See Garthoff, *supra* note 11, pp. 23–24 and 31.
14. Treaty Banning Nuclear Weapon Tests in the Atmosphere, in Outer Space and Under Water, *Arms Control Docs., supra* note 7, pp. 34–44. T.I.A.S. No. 5433; 2 U.S.T. 1313; 480 U.N.T.S. 43.
15. See, for example, Preparedness Investigating Subcommittee of the U.S. Senate Committee on Armed Services, "Hearings on the Military Aspects and Implications of Nuclear Test Ban Proposals," 88th Cong., 1st sess. (1963).
16. See Garthoff, *supra* note 11, pp. 25 and 35.
17. See 1962 G.A.Res. 1884 (XVIII), 18 U.N.G.A.O.R. Supp. 15 at 13, U.N. Doc. A/5515 (1963); Statement by Adlai Stevenson, U.S. representative to the United Nations on October 16, 1963, Arms Control and Disarmament Agency, *Documents on Disarmament* (hereinafter "ACDA Docs"), pp. 535, 537 (1963); and the statement made by the Soviet representative to the United Nations, found in 18 U.N.G.A.O.R. 18 (1963). These statements were preceded by a statement of the American position in early September. See Statement of Deputy Secretary of Defense Roswell Gilpatric, September 5, 1962, found in 108 Cong. Rec. S7007 (daily ed. 21 September 1962). (The United States would not deploy "bombs in orbit" if the Soviets refrained from doing so.) Soviet Foreign Minister Gromyko responded to the American position in an address to the UN General Assembly on 19 September 1963 by proposing an "agreement with the U.S. government to ban the placing into orbit of objects with nuclear weapons on board." *ACDA Docs.* at pp. 509, 523 (1963). President Kennedy agreed to the proposal in a UN General Assembly speech a day later. See idem, pp. 525, 528.
18. See, generally, Bernard A. Ramundo, *Peaceful Coexistence: International Law in the Building of Communism,* (Baltimore: Johns

Hopkins Univ. Press, 1967), pp. 43–71. In certain circumstances unilateral representations by national officials can give rise to an international legal obligation. See, for example, *The Nuclear Tests Cases,* 1974 I.C.J. 253, 457 digested in *American Journal of International Law* 69 (1975): 668. Compare Note, "Arms Control: SALT-Unilateral Policy Declaration by the U.S.," *Harvard International Law Journal* 19 (1978): 372 (hereinafter *HILJ*).

19. 22 U.S.C. §2573.
20. "Draft Declaration of the Basic Principles Governing the Activities of States in Exploration and Use of Outer Space," U.N. Doc. A/AC.105/C.1/L.6 (April 16, 1963). The Soviet Union also proposed (1) to ban surveillance from space as incompatible with mankind's loftier motives in conquering the cosmos, and (2) to require agreement before a country could implement space activities that could hinder in any way space exploration by other nations.
21. Resolution 1962 (XVIII). U.N. Doc. A/C.1/L.331 (1963). The text is reprinted in *Department of State Bulletin* 49 (1963): p. 1012.
22. See G.A.O.R. (XVIII), 1st Committee, 1342d meeting 2 December 1963, p. 161. Soviet international legal scholars continue to adhere to this view. See A.S. Piradov, ed., *International Space Law* (Moscow 1976): 259.
23. See *Department of State Bulletin* 47 (1962): p. 318.
24. American positions at the United Nations were developed following the submission of a report to NASA by the American Bar Foundation, which concluded that the United States was justified in using space for "non-aggressive military uses consistent with the... [U.N.] charter...." See generally Martin Menter, "Peaceful Uses of Outer Space and National Security," *International Lawyer* 17 (1983): 584–85.
25. See Jaksetic, *supra* note 5, at pp. 491–92; and Albert J. Esgain, "The Position of the United States and the Soviet Union on Treaty Law and Treaty Negotiations," *Military Law Review* 46 (1969): 35–36. See also Ramundo, *supra* note 18, pp. 60–64. Compare *International Space Law, supra* note 22, pp. 74–77.
26. See Paul G. Dembling and Daniel M. Arons, "The Evolution of the Outer Space Treaty," *Journal of Air Law and Commerce* 33 (1967): 425.
27. See *Department of State Bulletin* 54 (1966): 900.
28. The text of the Soviet response, which took the form of a letter to U.N. Secretary-General Waldheim, is reprinted by Dembling and Arons, *supra* note 26, p. 426 n. 32.
29. The U.S. draft, U.N. Doc. A/AC.105/32 (1966), is reproduced in *Department of State Bulletin* 55 (1966): 61. The Soviet draft, U.N.

Doc A/6352 (1966) is annexed (along with the American draft) to the "Report of the Legal Subcommittee on the Work of its Fifth Session" to COPUOS, U.N. Doc. No. A/AC.105/35. See Annex I at pp. 12–16 (Soviet Union) and pp. 6–9 (United States).

30. For a historical background to the development of the treaty, see Paul G. Dembling, "Treaty on Principles Governing the Activities of States in the Exploration and Use of Outer Space Including the Moon and Other Celestial Bodies," in N. Jasentuliyana and R.S.K. Lee, eds., 1 *Manual on Space Law,* (Dobbs Ferry, N.Y.: Oceana Publishers, 1979), pp. 1–52.
31. Resolution 2222 (XXI) (December 19, 1966). *Arms Control Docs., supra* note 7, pp. 48–56. T.I.A.S. No. 6347; 18 U.S.T. 2410, 610 U.N.T.S. 205.
32. The text of the UN charter is reprinted in Louis B. Sohn and Thomas Buergenthal, *Basic Documents on International Protection of Human Rights,* (New York: The Bobbs-Merrill Company, Inc., 1973), pp. 1–30.
33. See Myres Smith McDougal & Florentino P. Feliciano, *Law and Minimum World Public Order—The Legal Regulation of International Coercion* (New Haven: Yale University Press, 1961), pp. 231–41.
34. The UN General Assembly approved the definition of aggression as "...the use of armed force by a State against the sovereignty, territorial integrity or political independence of another State or in any manner inconsistent with the Charter of the United Nations as set out in this definition." G.A. Res. 3324(XXIX), 29 U.N. G.A.O.R., Supp. 142, U.N. Doc. A/9631 (1975).
35. See generally Dembling and Arons, *supra* note 26, pp. 447–51 for a description of the history of the negotiations concerning inspection.
36. See, for example, S.H. Lay & H.J. Taubenfeld, *The Law Relating to Activities of Man in Space,* (Chicago: Univ. of Chicago Press, 1970), pp. 97–102. See also Stein, "Legal Restraints in Modern Arms Control Agreements," *Am.J. Int'l L.* 66 (1972): 255, 262–64. The United States continues to rely upon this rationale with respect to antisatellite weapons. See U.S. Dept. of State, *Digest of U.S. Practices in International Law 1977,* (1979): 665–66. See also the statements of UN Ambassador Goldberg and Deputy Secretary of Defense Vance made in conjunction with the Senate's ratification of the Outer Space Treaty in U.S. Senate Committee on Foreign Relations, *Hearings on the Treaty on Outer Space,* 90th Cong., 1st sess., (1967): 23, 76, and 100.
37. Potential exceptions include high energy lasers now under development that rely upon nuclear explosions to produce the laser beams. Testing of these weapons in outer space would seemingly violate the Outer Space Treaty if the weapon completed at least one orbit, or, in

any event, the 1963 Limited Test Ban Treaty. See, generally, Gerard Smith, "A Dangerous Dream: Why Reagan's Plan Threatens the Nuclear Balance," *Baltimore Sun,* 29 March 1983. See also note 39 *infra*.

38. See, for example, Com. on the Peaceful Use of Outer Space, Legal Sub-Com., 5th Sess., 66th Meeting, 25 July 1966, U.N. Doc. A/AC 105/C.2/SR. 66 at 7 (1966). See also ibid. at 71st meeting, August 4, 1966, U.N. Doc. A/AC. 105/C.2/SR. 71 and Add. 1 at 8–9 (1966).
39. The Department of Defense has taken the position that "weapons that do not stay in space for one complete orbit are not considered to be 'in space'". See Everet Clark, "New Bomb Feared by U.S. Since 1966," *New York Times,* 4 November 1967. The question also arises whether "nuclear weapons" include nuclear-powered lasers or only traditional nuclear weapons capable of causing mass destruction. At the time the Outer Space Treaty was ratified, the American Legal Advisor to the Secretary of State publicly took the position that orbiting nuclear powered lasers and "small" nuclear weapons that would not cause mass destruction were both prohibited under Article 4 of the Outer Space Treaty. See "Loophole Seen in Space Treaty," *Science News* 19 (17 June 1967): 565–66.
40. T.I.A.S. No. 6599; 19 U.S.T. 7570; 672 U.N.T.S. 119.
41. *Arms Control Docs., supra* note 7, 109–11 and 158–59. T.I.A.S. No. 7186; 22 U.S.T. 1590 (1971); and T.I.A.S. No. 7654; 24 U.S.T. 1478 (1973).
42. Interim Agreement on Certain Measures with Respect to the Limitation of Strategic Arms, T.I.A.S. No. 7504; 23 U.S.T. 3462; Treaty on the Limitation of Anti-Ballistic Missile Systems, T.I.A.S. No. 7503; 23 U.S.T. 3435. The text of these treaties is also reprinted in *Arms Control Docs., supra* note 7, pp. 137–54. The June 1979 Treaty between the U.S. and U.S.S.R. on the Limitation of Strategic Offensive Arms (known as SALT II) ibid. at 201–34, contains the same provision on verification and is adhered to on a de facto basis because of the U.S. failure to ratify the Treaty, just as the Interim Agreement was after its expiration. See, generally, *HILJ, supra* note 18.
43. The text of President Reagan's speech was reprinted in the *New York Times,* 24 March 1983. Note that like many arms control agreements, the ABM treaty provides for withdrawal by a party on six months notice "[i]f it decides that extraordinary events related to the subject matter of this treaty have jeopardized its supreme interests."
44. See the description of existing weapons in the pamphlet entitled "Anti-Satellite Weapons: Arms Control or Arms Race?" (Union of Concerned Scientists, 30 June 1983): 9–12.
45. T.I.A.S. No. 7762; 24 U.S.T. 2389.

46. T.I.A.S. No. 8480; 28 U.S.T. 695.
47. T.I.A.S. No. 9614; 31 U.S.T. 333 (1977); See also U.N. Docs. A/Res/34/68 and A/34/664 (1979). For historical perspective and analysis of the Moon Treaty, see U.S. Senate Committee on Commerce, Science and Transportation, *Agreement Governing the Activities of States on the Moon and Other Celestial Bodies,* 96th Cong., 2nd sess. Committee Print, 1979.
48. For a discussion of the history of these provisions and assessment of their meaning, see Carl Q. Christol, "The Common Heritage of Mankind Provision in the 1979 Agreement Governing the Activities of States on the Moon and Other Celestial Bodies," *The International Lawyer* 14 (1980): 429. See also "Outer Space, International Law, International Regimes and The Common Heritage of Mankind," *J. Space L.* 10 (1982): 65–69; and the statement by Ambassador Richard W. Petree in "U.S. Statement in Special Political Committee, UN General Assembly, Excerpts on Moon Treaty, November 1, 1979," reprinted in *J. Space L.* 9 (1981): 161–63.
49. For a detailed discussion of remote sensing negotiations at the United Nations, see Gerald J. Mossinghoff and Laura D. Fuqua, "United Nations Principles on Remote Sensing: Report on Developments, 1970–1980," *J. Space L.* 8 (1980): 103. See also "Report of the Working Group on Remote Sensing," reprinted in *J. Space L.* 8 (1980): 70–75 as well as subsequent reports reprinted in *J. Space L.* 9 (1981): 121–26 and *J. Space L.* 10 (1982): 95–99. The text of draft principles governing the use by states of satellites for direct television broadcasting is also reprinted in *J. Space L.* 8 (1980): 187–92.
50. A translation of the Bogota Declaration may be found in *J. Space L.* 6 (1978): 193. See generally, Ivan A. Vlasic, "Disarmament Decade, Outer Space and International Law," *Revue de Droit de McGill* 26 (1981): 187–89; and the 1973 International Telecommunications Convention done at Malaga-Torremolinos. T.I.A.S. No. 8572, 28 U.S.T. 2495. The limits of ITU power to control GSO access and use, to issue sanctions because of harmful interference, or to regulate military radio installations are apparent in Articles 33, 35, and 38 of that convention.
51. See He Qizhi, "The Problem of Definition and Delimitation of Outer Space," *J. Space L.* 10 (1982): 162. See also Vladimir Kopal, "The Question of Defining Outer Space" *J. Space L.* 8 (1980): 154.
52. The Soviet Union first linked the effort to define outer space to establishment of a right of innocent passage in June 1978. See D. Goedhuis, "The Changing Legal Regime of Air and Outer Space," *International and Comparative Law Quarterly* 27 (1978): 593 n.18a. See also UN Doc. A/AC 105/PV. 185, p. 42.

53. See, for example, Stephen Gorove, "Arms Control in Outer Space," *J. Space L.* 10 (1982): 75 (Report on the COPUOS Legal Subcommittee meetings in February, 1982).
54. See UN Doc. CD/9 (March 26, 1979). See also Fernando Lay, "Space Law: A New Proposal," *J. Space L.* 8 (1980): 55–58.
55. Ibid., pp. 50–54.
56. See UN Doc. A/S-10/PV. 3 p. 16 (25 May 1978). See also the French memorandum explaining the function of the proposed agency. UN doc. A/S-10/AC. 1/7 (1 June 1978). See, generally UNGA Res. 33/71 (14 December 1978) and UNGA Res. 34/83 E. (11 December 1979).
57. See, for example, UN Doc. A/AC, 105/PV 220 (1981).
58. See UNGA 36/97C entitled "Preventing an Arms Race in Outer Space," UN Doc., A/Res/36/97 (9 December 1981). See also UNGA Res. 36/99 (15 January 1982).
59. The draft treaty is annexed to UNGA Res. 36/97 (15 January 1982) and is reprinted in *J. Space L.* 10 (1982): 27–29.
60. See D. Goedhuis, "Some Observations On the Efforts to Prevent A Military Escalation in Outer Space," *J. Space L.* 10 (1982): 21.
61. See Yash Pal, "Report on UNISPACE '82 and Beyond," *J. Space L.* 10 (1982): 181; and N. Jasentuliyana, "The Second United Nations Conference on the Exploration and Peaceful Uses of Outer Space," *J. Space L.* 10 (1982): 188; See also N. Jasentuliyana, "The Work of the United Nations Committee on the Peaceful Uses of Outer Space in 1982," *J. Space L.* 10 (1982): 43.
62. See UN Doc. A/Conf. 101/5 (1982).
63. See Jasentuliyana (note 61), p. 191.
64. See UNISPACE '82 Report, UN Doc. A/Conf. 101/10 at 141 ¶¶13–14. The recommendations of UNISPACE '82 are summarized in *J. Space L.* 10 (1982): 237–49.
65. See, for example, UNGA Res. 37/92 (10 December 1982) on "Preparation of an International Convention on Principles Governing the Use by States of Artificial Earth Satellites for Direct Television Broadcasting," reprinted in *J. Space L.* 10 (1982): 252–55.
66. Clarence A. Robinson, Jr., "Panel Urges Defense Technology Advances," *Aviation Week & Space Technology* (17 October 1983): 16–18.
67. "Reagan Reported To Agree On Plan to Repel Missiles," *New York Times,* 1 December 1983.
68. 97 Stat. 614; P.L. 98-94, 98th Cong., §1235.
69. While it is not clear whether the phrase "in space" was used in the manner employed by the Department of Defense with respect to the Outer Space Treaty in 1967 (see note 39, supra) the legislative debate about the amendment seems to indicate that Congress was not as

precise. See Cong. Rec. S10262 (18 July 1983) (remarks of Senator Warner). In any event, the statute does not prohibit tests against unoccupied points in space.

70. See H.J. Res. 120, 98th Cong., 1st sess. (1983); S.J. Res. 129, 98th Cong., 1st sess. (1983). See also Senate Committee on Foreign Relations Report No. 98-342 entitled *Outer Space Arms Control Negotiations* (18 November 1983), recommending S.J. Res. 129, as amended.
71. See *Dangerous Stalemate: Superpower Relations in Autumn 1983, A Report of a Delegation of Eight Senators to the Soviet Union,* S. Doc. 98-16, 1st sess. (22 September 1983): 13–14, 27–28 ("Pell Report").
72. The text of the treaty (unofficial Novosti translation) is reprinted in the Pell Report, pp. 35–37. For a more thorough analysis, see Chapters 8 and 9, infra.

REFERENCES

Akehurst, Michael. "Custom as a Source of International Law." *1974–75 British Yearbook of International Law.* Oxford: Clarendon Press, 1977.

Arms Control and Disarmament Act of 1961. 22 United States Code sec. 2573.

"Arms Control: SALT–Unilateral Policy Declaration by the U.S." (Discussion of a letter by Secretary of State Cyrus Vance to Sen. John Sparkman of 21 September 1977 regarding U.S. compliance with SALT I Interim Agreement.) *Harvard International Law Journal* 19 (1978): 372.

Bechhoffer, Bernard G. *Postwar Negotiations for Arms Control.* Washington, D.C.: The Brookings Institution, 1961.

Bourley, Michael. "The Contributions Made by International Organizations to the Formation of Space Law." *Journal of Space Law* 10 (1982).

Brennan, Donald G. "Arms Control in Outer Space." In L. Bloomfield, ed., *Outer Space–Prospects for Man and Society.* Englewood Cliffs, N.J.: Prentice-Hall, 1962.

Christol, Carl Q. "The Common Heritage of Mankind Provision in the 1979 Agreement Governing the Activities of States on the Moon and Other Celestial Bodies." *The International Lawyer* 14 (1980): 429.

———. "Outer Space, International Law, International Regimes and the Common Heritage of Mankind." *Journal of Space Law* 10 (1982): 65–69.

Clark Everet. "New Bomb Feared by U.S. Since 1966." *New York Times,* 4 November 1967.

Dembling, Paul G. "Treaty on Principles Governing the Activities of States in the Exploration and Use of Outer Space Including the Moon and Other

Celestial Bodies." In N. Jasentuliyana and R.S.K. Lee, eds., *Manual on Space Law*. 4 vols. Dobbs Ferry, N.Y.: Oceana Publishers, 1979.

———, and Daniel M. Arons."The Evolution of the Outer Space Treaty." *Journal of Air Law and Commerce* 33 (1967).

Department of Defense Authorization Act, 1983. 97 Stat. 614. Public Law 98-94, sec. 1235. (Tsongas amendment on antisatellite weapon tests.)

Esgain, Albert J. "The Position of the United States and the Soviet Union on Treaty Law and Treaty Negotiations." *Military Law Review* 46 (1969).

Galloway, Eilene M. "Consensus Decision-Making by the United Nations Committee on the Peaceful Uses of Outer Space." *Journal of Space Law* 7 (1979).

Garthoff, Raymond L. "Banning the Bomb in Outer Space." *International Security* 5 (Winter 1980–81).

Goedhuis, D. "The Changing Legal Regime of Air and Outer Space." *International and Comparative Law Quarterly* 27 (1978): 588–92.

———. "Some Observations on the Efforts to Prevent a Military Escalation in Outer Space." *Journal of Space Law* 10 (1982): 13–30.

Gorove, Stephen. "Arms Control in Outer Space." Report on the Activities of the Legal Subcommittee of COPUOS. *Journal of Space Law* 10 (1982): 74–75.

He Qizhi. "The Problem of Definition and Delimitation of Outer Space." *Journal of Space Law* 10 (1982): 157–64.

International Court of Justice. *Nuclear Test Cases*. 1974 I.C.J. 253, 457. (Digested in *American Journal of International Law* 69 (1975): 668–83.)

Jaksetic, Emilio. "The Peaceful Uses of Outer Space: Soviet Views." *American University Law Review* 28 (1979).

Jasentuliyana, N. "The Work of the United Nations Committee on the Peaceful Uses of Outer Space in 1982." *Journal of Space Law* 10 (1982): 41–46.

———, and R.S.K. Lee, eds. *Manual on Space Law*. 4 vols. Dobbs Ferry, N.Y.: Oceana Publishers, 1979–81.

Kopal, Vladimir. "The Question of Defining Outer Space." *Journal of Space Law* 8 (1980): 154–73.

Lay, Fernando. "Space Law: A New Proposal." *Journal of Space Law* 8 (1980): 41–57.

Lay, S.H., and H.J. Taubenfeld. *The Law Relating to Activities of Man in Space*. Chicago: University of Chicago Press, 1970.

"Loophole Seen in Space Treaty." *Science News* 19 (17 June 1967).

McDougal, Myres S. and Florentino P. Feliciano. *Law and Minimum World Order—The Legal Regulation of International Coercion*. New Haven: Yale University Press, 1961.

Menter, Martin. "Peaceful Uses of Outer Space and National Security." *International Lawyer* 17 (1983).

Mossinghoff, Gerald J. and Laura D. Fuqua. "United Nations Principles on Remote Sensing: Report on Developments, 1970–1980." *Journal of Space Law* 8 (1980): 103–53.

Pal, Yash. "Report on UNISPACE '82 and Beyond." *Journal of Space Law* 10 (1982): 181–86.

Permanent Court of International Justice. *S. S. Lotus.* 1927 Ser. A., No. 10.

Piradov, A.S., ed. *International Space Law.* Moscow: Progress Publishers, 1976.

"President Reagan's Speech on Military Spending and a New Defense." *New York Times,* 24 March 1983.

Ramundo, Bernard A. *Peaceful Coexistence: International Law in the Building of Communism.* Baltimore: Johns Hopkins University Press, 1967.

Robinson, Clarence. "Panel Urges Defense Technology Advances." *Aviation Week and Space Technology* (17 October 1983): 16–18.

Smith, Gerard. "A Dangerous Dream: Why Reagan's Plan Threatens the Nuclear Balance." *Baltimore Sun,* 29 March 1983.

Sohn, Louis B., and Thomas Buergenthal. *Basic Documents on International Protection of Human Rights.* New York: Bobbs-Merrill, 1973.

Stein, E. "Legal Restraints in Modern Arms Control Agreements." *American Journal of International Law* 66 (1972).

Union of Concerned Scientists. "Anti-Satellite Weapons: Arms Control or Arms Race?" Cambridge, Mass., June 30, 1983. Pamphlet.

United Nations. Ad Hoc Committee on the Peaceful Uses of Outer Space. Report. A/4141/25 (1959).

United Nations Committee on the Peaceful Uses of Outer Space (COPUOS). "Draft Declaration of the Basic Principles Governing the Activities of States in Exploration and Use of Outer Space." A/AC.105/C.1/L.6 (16 April 1963).

United Nations Committee on Disarmament. CD/9 (26 March 1979). (Protocol to the Outer Space Treaty proposed by Italy.)

———. 36th Session. "Draft Treaty on the Stationing of Weapons of Any Kind in Outer Space." (USSR) Annexed to A/Res/36/99 (15 January 1982) and reprinted in *Journal of Space Law* 10 (1982): 27–29.

———. Second U.N. Conference on the Exploration and Peaceful Uses of Outer Space. Report. A/Conf.101/10 (1982).

———. First Committee. Verbatim Official Record. A/C.1/L.219 (1958). (Soviet proposal to ban the military use of outer space.)

———. First Committee. "Declaration of Legal Principles Governing the Activities of States in the Exploration and Use of Space." G.A. Res. 1962 (XVIII). A/C.1/L.331 (1963).

———. A/S-10/PV.3, p. 16 (25 May 1978). (French proposal for an International Satellite Monitoring Agency, explained in A/S-10/AC.1/7 [1 June 1978].)

United Nations. Secretariat. Office of Legal Affairs. E/CN.4/L.610 (2 April 1962). (On the nonbinding nature of UN resolutions.)

U.S. Arms Control and Disarmament Agency. *Arms Control and Disarmament Agreements.* Washington, D.C.: GPO, 1980.

———. *Documents on Disarmament.* Washington, D.C.: GPO, 1962– .

U.S. Congress. Senate. "Dangerous Stalemate: Superpower Relations in Autumn 1983, A Report of a Delegation of Eight Senators to the Soviet Union." S. Doc. 98-16. 98th Cong., 1st sess. 22 September 1983.

U.S. Congress. Senate. Committee on Aeronautical and Space Sciences. *Statements by Presidents of the United States on International Cooperation in Space—A Chronology: October 1957–August 1971.* S.Doc. No. 92-40. 92nd cong., 1st sess. 1971.

———. Committee on Armed Services. Preparedness Investigating Subcommittee. *Hearings on the Military Aspects and Implications of Nuclear Test Ban Proposals.* 88th Cong., 1st sess. 1963.

———. Committee on Foreign Relations. *Hearings on the Treaty on Outer Space.* 90th Cong., 1st sess. 1967.

———. Committee on Foreign Relations. *Outer Space Arms Control Negotiations.* Report on S. J. Res. 129, as amended. SFRC Report No. 98-342. 98th Cong., 1st sess. 18 November 1983.

———. Committee on Commerce Science, and Transportation. *Hearings, Agreement Governing the Activities of States on the Moon and Other Celestial Bodies.* 96th Cong., 2d sess. Committee Print. 1979.

———. Congressional Record 108 (21 September 1962). Statement of Deputy Secretary of Defense Roswell Gilpatric, 5 September 1962.

U.S. Department of State. *Digest of U.S. Practices in International Law,* 1977. (1979).

U.S. Department of State. *United States Treaties and Other International Agreements Series* (T.I.A.S.).

———. "International Telecommunications Convention (Malaga-Torremolinos, 1973)." T.I.A.S. No. 8572. 28 U.S.T. 2495.

———. "Chicago Convention on International Civil Aviation." T.I.A.S. No. 1591 (1944). 61 Stat. 1180.

———. "Treaty on the Return of Astronauts and Objects Launched into Space." T.I.A.S. No. 7503 (1968). 19 U.S.T. 7570.

———. "Convention on International Liability for Damage Caused by Space Objects." T.I.A.S. No. 7762 (1973). 24 U.S.T. 2389.

———. "Convention on Registration of Objects Launched into Outer Space." T.I.A.S. No. 8480 (1975). 28 U.S.T. 695.

"U.S. Statement in Special Political Committee, U.N. General Assembly, Excerpts on Moon Treaty, November 1, 1979." (Statement of U.S. Ambassador Petree.) Reprinted in *Journal of Space Law* 9 (1981): 161.

Vlasic, Ivan. "Disarmament Decade, Outer Space and International Law." *Revue de Droit de McGill* 26 (1981):

8 SOVIET LEGAL VIEWS ON MILITARY SPACE ACTIVITIES

Malcolm Russell

Since the launching of Sputnik in 1957, Western countries have amassed a large amount of information on Soviet military activities in space. In recent years, analysts have pieced together important data about Soviet reconnaissance, navigation, and communication satellites, and about various space weapon programs. Less time, however, has been spent assessing Soviet policy on the use of space for military purposes. Given the rapidly evolving military situation in space and the complexity of the issues, an examination of Soviet views appears timely.

Perhaps the most extensive Soviet treatment of military activities in space can be found in Soviet legal publications. Books, pamphlets, and journal articles by Soviet international law specialists devote much more attention to issues of space policy than do comparable works by foreign policy specialists or military experts. The technical nature of many space issues, as well as their lesser importance relative to subjects such as nuclear weapons has generally discouraged extensive discussion by foreign policy commentators in the Soviet Union. At the same time, the controversial nature of space military activities and the absence of a consensus on the importance of such missions

Revised version of an article published in the *Harvard International Law Journal* 24 (Winter 1984).

have kept Soviet military publications rather quiet on the subject. To date, only a handful of articles from the Soviet military press have addressed, even tangentially, the military uses of space.

By contrast, the entire range of human activities in space has consistently engaged the attention of Soviet international legal experts. In addition to maintaining a deep academic interest in the subject, Soviet specialists in international law have spearheaded an effort by the Soviet Union to inject favored socialist international legal principles into the young and mutable field of space law. As with the Law of the Sea, these authorities have played an important role in formulating Soviet policy and participating in negotiations with the West. To take one notable example, the head of the Soviet delegation to the Soviet-American talks on antisatellite weapons in 1978–79 served for many years as deputy chief of the legal department of the Ministry of Foreign Affairs and was the author of several articles on space and international law.[1]

For a number of reasons, Soviet legal sources can prove useful in trying to ascertain Soviet views on space military activities. Much of the literature serves a limited Soviet and foreign academic readership rather than a mass international audience. While all Soviet publications have a high propaganda content and much recycled criticism from the Western press, it is possible to find a good deal of legal writing that is relatively nonideological. The greater detail and precision of work by specialists like lawyers often provide useful clues to the purposely ambiguous and politicized aspects of Soviet policy. Indeed, the reductionist quality of law—enunciating starkly the legality or illegality of a certain activity—can introduce some clarity into Soviet views, including policy shifts. Although such legal writing may ultimately have limited predictive value concerning future Soviet conduct—attitudes may, after all, change rapidly or not be disclosed, depending on the circumstances—they can illuminate some of the more fundamental features of Soviet space policy, as well as potential areas of concern to Western policymakers.

THE NATURE OF SOVIET INTERNATIONAL LAW

It is impossible to evaluate Soviet legal views on space without considering their relationship to Soviet international law and its special mission. In the Soviet view, international law cannot be divorced from the class struggle. Since, in Marxist terms, international relations compose part of the superstructure of society, relations between states with different social systems constitute class relations. "The ideological foundation of a state's foreign policy is the definite system of principles and views of the ruling class of that state."[2] As law also represents a class phenomenon, so it becomes a weapon in the ongoing struggle for world socialism.

In the ideological context of building communism, the Soviets unabashedly place international law in the service of the master builder, the Soviet state. As articulated by the leading modern Soviet international jurist, Grigorii Tunkin, "the international legal position of a state is determined by the basic principles of its foreign policy."[3] In one sense, this formulation differs only in its degree of candor from that of other modern nations. In the Soviet view, however, this standard does not amount to pure self-interest, since by definition the world Socialist movement has an international, rather than bourgeois national, character.[4]

Though long viewed as an offensive, "provocative" weapon in Soviet foreign policy, within the last few decades international law has also effectively played a defensive or "ordering" role in the Soviet Union's relations with other states. As such, the Soviet Union has engaged in a delicate balancing act, pressing its own "provocative" interpretations of legal rules as far as possible without endangering the shared understanding of various principles that ultimately ensures order for the socialist community.[5] At the risk of disrupting the reciprocal, patterned behavior on which international—and Soviet—stability depends, the Soviets have therefore "incorporated" international law but endeavored to inject it with broad new meanings advantageous to the Soviet Union.[6]

The Soviet Union has advanced its new approach to international law under the umbrella of "peaceful coexistence." Since Khrushchev first enunciated the principle in 1956, peaceful coexistence has variously described a foreign policy goal, international relations in the nuclear age, and a juridical concept embracing the fundamental tenets of Soviet international law.[7] In the last case, Soviet international jurists have effectively repackaged several fundamental principles of international law—including nonintervention in the internal affairs of states, respect for territorial integrity, the equality of states—and made them the centerpiece of a new socialist international law. Although such ambiguous phrases ostensibly say nothing new about the basic relations between states, they may acquire slightly new meanings under the progressive influence of socialist usage. For example, notions about the equality of states have traditionally revolved around the need for a state to respect the interests of other nations in pursuing a course of action in the international arena. Over time, the Soviets have expanded that idea and tended to invoke it in any situation where a foreign state has gained a unilateral advantage of whatever kind. Consistent with the Soviet view that agreements between states with different social systems in no way diminish the ideological struggle, peaceful coexistence—and its more recent analogue, detente—demonstrate that East and West share certain principles only up to a point.

Promotion of the peaceful coexistence principles in fact reflects a rather conservative, defensive posture and has led the Soviets to favor the conclusion of treaties as a means of defining the international legal obligations of the Soviet Union.[9] The Soviets' penchant for predictability in relations with other states tends to make them highly legalistic in defining the obligations of states-parties under an agreement.[10] Perhaps for that reason, the Soviets have found the Standing Consultative Commission (SCC) established under SALT I a congenial place in which to seek clarification and pin down American positions. By all accounts, the SCC has greatly aided SALT I compliance by allowing the United States and the Soviet Union to lodge complaints in a low-key setting and by satisfying the Soviets' need, as a matter of pride, to answer American demands with corresponding challenges of their own.[11]

This fundamental desire for predictability and control in international relations finds ample expression in the Soviet approach to space law. Because the ideological struggle also informs the competition between East and West in space, Soviet legal theoreticians reject the idea of space as a "legal vacuum" and condemn those "who would retain a free hand" to disrupt socialist space efforts.[12] "Freedom of space," warns one legal expert, "should not be used as a pretext for violating sovereign rights on earth."[13] According to a Soviet text on space law, the nature of the modern nation-state system and the mutual hostility of the capitalist and socialist camps require the application of international order to space. Consequently, Soviet jurists promote the relevance of peaceful coexistence to space law by stressing the earth-oriented character of space activities.[14]

But the Soviets do more than merely assert the need for space law to comport with general precepts of international law. They affirm the potential applicability of the latter to any situation involving the former. Thus, Soviet legal authorities regard the broadcasting of hostile propaganda from space to a foreign country as involving both space law and general questions of territorial integrity, whereas many Western international lawyers treat it, if at all, only as a subject of telecommunications law. This tendency to address the legality of space activities by referring reflexively to general international legal principles enables the Soviets to change the focus of many legal debates. By relying on the Socialist principles of peaceful coexistence, such as respect for the equality of states or state sovereignty, the Soviet Union can effectively denounce a wide range of international conduct as inimical to vital Soviet interests. Regardless of the actual legal effect of such statements, their cumulative impact on the political perceptions of other states may be great.

SPACE LAW AS IDEOLOGY: THE QUESTION OF "PEACEFUL PURPOSES"

For years, one of the central precepts of Soviet space law has been the ideal of preserving the use of outer space for peaceful purposes. As with the concept of peaceful coexistence, the expression "peaceful purposes" carries a specific meaning for Soviet jurists. While Western legal scholars have held that a peaceful purpose describes any nonaggressive activity, their Soviet counterparts have consistently chosen to define the term as any "nonmilitary" conduct.[15] While bolstering the peace-loving image of the Soviet Union, the peaceful purposes formula has undergone numerous modifications. It has long ceased to describe accurately Soviet space policy, since fully two-thirds to three-fourths of all Soviet spaceflights have served special military functions.[16] Nevertheless, it continues to operate as an aspirational goal of Soviet space law and to generate substantial political capital for the Soviet Union in the ongoing ideological struggle with the West.

In the late 1950s and early 1960s, the Soviet Union closely identified the use of space for peaceful purposes with its plan for general and complete disarmament (GCD). In effect, the Soviets proposed to "neutralize" space by swapping their supposed military superiority in that realm—in particular, intercontinental ballistic missiles—for an

American withdrawal of missiles and bombers from overseas bases near the Soviet Union.[17] Even as the Soviet Union advanced this all-or-nothing proposition, Khrushchev did not hesitate to make bellicose threats about the possible use of space to bombard the United States, and Soviet jurists sought to justify the military use of space for self-defense contingencies.[18] "No organic contradiction exists," pronounced two legal experts dialectically, "between the use of space for scientific purposes and its use for the protection of national security."[19] Even less in harmony with their stated goals, the Soviets proceeded to test total and fractional orbital bombardment systems through the larger part of the decade.

So long as no positive law governed outer space, the Soviet Union was free to champion a highly idealized definition of peaceful purposes. When it signed the 1967 Outer Space Treaty, however, the Soviet Union was forced to bring its interpretation into line with legal reality. According to Article 4(2) of the treaty, "peaceful purposes" applies only to activities on the moon and other celestial bodies, not to operations in outer space generally. Thus, even under the Soviet interpretation, space military activities are permitted in the latter by implication. Soviet jurists thus faced a choice: They could drop their own definition and accept that of their Western colleagues, or they could continue to call for an ideal state of affairs at odds with both their new reconnaissance capability and the Outer Space Treaty.

As it turned out, Soviet legal experts shrewdly sidestepped the issue. They held on to the definition of peaceful purposes as describing nonmilitary activities but, consistent with the Outer Space Treaty, applied it only to operations on the moon and celestial bodies.[20] At the same time, using different terminology and less strident rhetoric, they continued to support a "demilitarized" or even "neutralized" outer space for the future. A neutralized outer space would preclude its use even in wartime and apparently would prohibit ballistic missiles from passing through space.[21] As with earlier proposals of this nature, the neutralization of space would go hand in hand with general and complete disarmament.

Despite such rhetorical continuity, the Soviet Union has concurred with the West on the legal status of military activities in space. "Hardly disputable now," concedes one Soviet jurist, "is the thesis that the Space Treaty has not fully banned any military activity."[22] The leading Soviet treatise on space law further states that "launching military objects into outer space has not yet been forbidden, nor has the use of outer space for military maneuvers or for testing various types of weapons."[23] Indeed, only on the subject of military operations on the moon and celestial bodies do Soviet international lawyers clash with many Western theorists. In that context, the Soviet interpretation of a "peaceful purpose" as a "nonmilitary" activity may perhaps more closely approximate the intent of the Outer Space and Moon Treaties than a "nonaggressive" definition. As pointed out by several scholars, the prevailing Western view might allow defensive or deterrent weapons on celestial bodies, thus clashing with the spirit of the treaties and their strict ban on the establishment of military installations.

Alongside prescriptive generalities about turning space into a peaceful sanctuary, Soviet jurists have gradually developed a number of more detailed and descriptive legal views. In the 1960s, Soviet legal experts did little more than paraphrase arguments for general and complete disarmament and for a threatening military capability. Today, as the next section of this chapter demonstrates, the same authorities use a more sophisticated vocabulary and appear more willing to propose specific changes in international law. As a result, the Soviet Union has been more flexible in suggesting international legal arrangements that fall short of an ultimate solution. While Soviet jurists continue to link the demilitarization of space with universal disarmament, they do not preclude the "possibility of signing international treaties forbidding or limiting specific aspects of the military use of space."[24] During negotiations in 1969–71 on the banning of nuclear weapons on the seabed, the Soviet Union initially called for the complete demilitarization of the ocean floor but subsequently adopted the narrow American position prohibiting only nuclear weapons.[25] Despite the countries' different strengths and weaknesses in space, it is likely that the Soviet Union could strike a similarly pragmatic bargain with respect to that environment as well.

SOVIET VIEWS ON SPECIFIC MILITARY SPACE ISSUES

Apart from news articles that simply parrot critical stories in the Western press, Soviet discussion of space military activities has focused with remarkable consistency on five discrete topics: self-defense in space, satellite reconnaissance, direct television broadcasting by satellite, orbital rendezvous, and the space shuttle. Listed in the order they have been addressed by the Soviets, these subjects receive extensive attention in Soviet legal sources and haphazard treatment in news and foreign policy publications. Meanwhile, only satellite reconnaissance has merited study in military journals.

Certain subjects are conspicuous by their absence. For one thing, the Soviets have barely broached the subject of antisatellite weapons (ASATs). They have never admitted that they have such weapons, even at the Soviet–American ASAT talks in 1978–79.[26] Even references to the American ASAT, scheduled for deployment in the near future, remain scarce. Although a 4 April 1977 TASS radio broadcast referred to "a secret American program to create a means of destroying enemy satellites,"[27] the ASAT talks revealed that neither country considered ASATs banned by the terms of the treaty, and further arguments of this kind have all but disappeared. Similarly missing, until very recently, has been serious commentary on the relative merits of a ban on the use of force in space, as opposed to a ban on specific weaponry. American policymakers have publicly grappled with that problem for several years now. Finally, such subjects as verification and possible negotiating positions turn up but rarely in Soviet publications. Information about any of these topics can likely be obtained, if at all, only from Soviet pronouncements at the ASAT talks and from the terms of the 1981 and 1983 Soviet draft treaties banning weapons from space.

In the meantime, the Soviet press has become increasingly militant in response to the Reagan administration's ambitious space plans. In a well-honed litany reminiscent of Soviet writing in the 1950s and early 1960s, several Soviet publications have featured articles on themes like "American Efforts to Achieve Military Supremacy in Space" and "Pentagon Battle Plans for Outer Space." The military journal *Krasnaia zvezda* (Red Star) has opined that "the Pentagon's unrestrained desire to militarize space...fits in perfectly" with an "American goal to achieve a first-strike capability over the USSR."[28] Many stories have also dealt with the increasingly large American research and development program for space military activities, as well as the variety of possible space weapons which the United States could deploy. However, only articles that discuss the five major subject areas treat Soviet space military policy in any depth.

Self-defense in Space

East and West both share the view that states have the same right to exercise self-defense in space that they do on earth. They differ, however, in their interpretation of that right. Article 51 of the United Nations Charter permits "individual or collective self-defense if an armed attack occurs against a Member of the United Nations." Article 2(4) of the Charter simultaneously calls upon states to refrain "from the threat or use of force." The Soviets have always gone further, however, and since the 1930s have sought a legal basis for self-defense that would identify as an aggressor the state that first takes hostile action.[29] Most Western states have opposed this standard and insisted that an act only be labeled aggressive if it was undertaken to achieve some prohibited objective.[30]

Following a concerted lobbying effort, the so-called Soviet "priority" formulation gained qualified acceptance in 1974 with the adoption by the United Nations of a Consensus Definition of Aggression.[32] That document defines aggression simply as the use of armed force in any manner inconsistent with the UN Charter and declares that "the first use of armed force by a state in contravention of the Charter shall constitute *prima facie* evidence of an act of aggression." As a feeble concession to Western states, the Consensus Definition also allows the UN Security Council to refrain from branding an act as aggressive if circumstances do not merit it. Despite loud objections from American international lawyers, the Consensus Definition has continued to gain adherents in the world community.

The Soviets have claimed that their first-in-time definition of aggression in outer space and elsewhere acts as a necessary counterweight to the popular Western concept of anticipatory self-defense, under which states may take defensive action in anticipation of an imminent threat of armed attack.[32] Soviet space law experts insist that only an actual armed attack may justify a state's taking self-defense measures. "Use of force in regard to foreign space objects based only on suspicion of another side's intentions," stated one Soviet jurist recently, "would flagrantly contradict...the right to self-defense as proclaimed in the U.N. Charter."[33] In order to avoid the "legalization of anarchy" and stay one step ahead of its capitalist enemies, the Soviet Union implicitly subscribes to the "priority" definition of aggression in space operations.[34]

It is difficult to distinguish what the Soviets regard as legitimate self-defense. Indeed, the Soviets' definition of aggression apparently gives them precisely the kind of subjective discretion to initiate self-defense that they discern in most Western legal writing. Actually, the West generally has not sought to apply anticipatory self-defense notions, especially in to the strategic arena, but has relied on the threat of second-strike retaliation as a deterrent to Soviet attack. The Soviet Union, on the other hand, has posited that a major conflict between East and West will arise from a deepening crisis; that the Soviet Union will have some warning of an impending war. Whether or not the West in fact strikes first, the Soviets believe they will know with certainty that war is inevitable, and at that time an overwhelming preemptive strike will be unleashed.[35]

Because Soviet self-defense doctrines have served largely unchanged for over two decades as verbal deterrents in support of abstract strategic doctrine, their relevance to more limited East-West confrontations in space is probably somewhat limited. In general, it is unlikely that the Soviet Union would be any more "trigger-happy" than the United States in defending itself against possible threats of attack. The potentially grave consequences of an attack against any space object of a foreign country, particularly a military satellite, suggest that a country will only take such an action when faced with a truly imminent and significant threat. The Soviet Union's emphasis on state sovereignty and its hypersensitivity to foreign intelligence gathering would lead one to expect a low threshold of tolerance for satellite interlopers. The evolution of its official attitude toward space-based reconnaissance, however, suggests a somewhat different result.

Satellite Reconnaissance

Soviet views on the legality of satellite reconnaissance have undergone substantial change over the years. As the first American satellites went into operation in the early 1960s, Soviet jurists made direct comparisons between space and aircraft reconnaissance. In 1961, with the downing of Gary Powers's U-2 fresh in mind, the leading Soviet space law authority argued that "from the viewpoint of the security of a state, it makes absolutely no difference from what altitude espionage over its territory is conducted."[36] The Soviets never indicated clearly whether the intrusion into sovereign Soviet space or the nature of the data collected by such reconnaissance constituted the more grievous violation of international law. Since, however, the oft-stated concern in the early years of the space age was that reconnaissance information would facilitate an American first-strike, Soviet jurists asserted that "the right to destroy...any space ship threatening the security of the state is undeniable."[37]

With the development of Soviet reconnaissance satellites and the abandonment of sovereign claims to any part of outer space came signs of a change in Soviet attitudes. As early as 1964, a Soviet space law expert noted the usefulness of monitoring compliance with the Test Ban Treaty from space.[38] Four years later, a mid-level officer wrote in the influential military journal *Voennaia mysl* (Military Thought) that satellite overflights could no longer constitute a violation of sovereignty. The officer doubted whether such arguments would have any future validity "inasmuch as an artificial earth satellite is in essence a global object and cannot but go beyond the boundaries where it was launched."[39] Finally in 1972, a prominent arms control commentator candidly acknowledged the verification provisions of the SALT I agreements and remarked on how satellites "made it considerably easier to reach agreement since [they] removed the question of conducting on-the-spot inspections, which had been a stumbling block previously."[40]

Even as the Soviets softened their position on satellite surveillance for certain purposes, much Soviet rhetoric remained hostile to the practice. The general Soviet press still made passing reference to "space espionage" in the mid-1970s,[41] while a *Krasnaia zvezda* article several months after the signing of the SALT I agreements still seemed to suggest that intelligence-gathering satellites might carry out impermissible operations.[42] Even in 1979 the same publication referred to United States "spy satellites" and condemned American space "espionage" programs.[43] The Soviets seemed uneasy about acquiescing in an area of activity where American technology still held sway.

Space law experts have tried to reconcile ambivalent Soviet attitudes toward satellite reconnaissance by placing limits on legitimate surveillance. Because each involves important questions of state sovereignty, the existence and scope of satellite reconnaissance must, say the experts, be ratified by agreement or otherwise. Soviet jurists accordingly view SALT I as having legitimized that degree of surveillance necessary to monitor compliance with the treaty. "The volume of monitoring [*kontrol*] or the competence of the monitoring body or the methods of inspection" must not "extend beyond what is objectively needed to observe the fulfillment of the agreement."[44] Using satellites for objectives other than verifying treaty adherence constitutes "a specific type of interference in the internal defense affairs of other states."[45]

It appears impossible to apply the Soviet standard in practice or to lend it credence as other than an ideological statement of principle. Existing treaties do not specify the proper scope of surveillance. There does not seem to be a way practically to separate permissible "routine" monitoring data from impermissible "incidental" intelligence. Since SALT I appears to permit surveillance of the entire territory of each of the parties, even arguments for geographic limitations on reconnaissance have little relevance.[46] Most likely the Soviet Union continues to hedge on full acceptance of satellite surveillance for traditional reasons of secrecy and sovereignty. Though it is hard to think of anything beyond the vision of today's high-resolution satellites, the Soviet Union would possibly still like to have the flexibility to object to future surveillance technologies and programs. Insisting on surveillance "by agreement" further ensures that future developments are discussed—as with SALT I—on the familiar, exclusive, and confidential state-to-state basis with which the Soviet Union is comfortable.[47]

The Soviet position on remote sensing, an analogous activity involving the collection of terrestrial environmental data by satellites, sheds some light on the Soviet approach to satellite reconnaissance. For many years now, the Soviets have taken the stand that the collection of data on the natural resources of a country and their use by other countries or private enterprise represents a violation of state sovereignty. Accordingly, the Soviets and several Third World nations have insisted that dissemination of some of the "sense" data—those with ground resolution of fifty meters or better—must depend on the consent of the affected state.[48] Notwithstanding the commercial availability of such data from the Landsat satellite, the Soviet Union has derived substantial political benefits from its stand and has forced extensive international negotiations on the subject in the United Nations.

In sum, it appears that Soviet denunciation of military satellite reconnaissance does not merit serious Western attention. Although at least one Soviet commentator suggests that the Soviet Union could take "defensive" countermeasures to deal with a serious reconnaissance threat to national security, such a statement seems like mere bluster.[49] The Soviets have explicitly acknowledged the noninterference clauses of SALT I and II and most likely would take any complaints they had to the Standing Consultative Commission. Under an expansive reading of Article 9 of the Outer Space Treaty, the Soviet Union could conceivably also demand consultations with a foreign state or discussions at the United Nations concerning any "possibly harmful interference" that might result from surveillance.[50] Thus far no hostile incidents have sprung from space-based foreign reconnaissance of the Soviet Union.[51]

Direct Television Broadcasting (DTB)

Although the broadcasting of American situation comedies from satellites may not seem an aggressive activity to Western observers, Soviet statements describe DTB, scheduled to begin operation in the United States in late 1984, as a serious potential threat to national security. Countless Soviet publications emphasize that "a state is entitled to decide for itself what information may be supplied to its population" and that consent of the receiving state should be obtained before satellite broadcasts commence.[52] Such protection is needed, according to Soviet writers, because "international mass information has at the present stage become one of the most powerful weapons in the ideological struggle of states."[53]

The Soviets also invoke UN General Assembly Resolution 110, adopted in 1947, which condemns "propaganda designed or likely to provoke or encourage any threat to the peace, breach of the peace or act of aggression." Soviet international lawyers have long interpreted Resolution 110 broadly to prohibit radio programs "propagandizing war, national and racial hatred, and other forms of animosity between peoples."[54] It is probably not inaccurate to say that any television or radio broadcasts not in accord with Soviet policy by definition promote such animosity.

The first definitive statement of Soviet policy on DTB emerged in 1972 in the form of a Soviet Draft International Convention on Principles of Direct Television Broadcasting. The proposal, which called for the equal rights of all states to broadcast from satellites "in the interest of peace, progress, and the development of mutual understanding," nevertheless insisted that such broadcasts be made only with the "express consent" of the receiving state. Most ominously, Article 9 of the Draft Convention indicated that a state could "use the means within its disposal of counter-acting DTB beamed at this state, not only [within] its territory, but also in outer space and other places beyond the bounds of national jurisdiction of any state."[55]

The Soviet Union modified its proposal in 1974 with a new set of working "Principles." The new document addressed the problem of overspill—unintended broadcasting of programs to peripheral areas of an adjacent country—by suggesting mandatory consultations between the sending and receiving states. In an effort to soften their stand before negotiation of the 1975 Helsinki accords, the Soviets also replaced the provision allowing unrestricted countermeasures with a stipulation that a state could use "measures which are recognized as legal under international law."[56] Peaceful coexistence principles and Soviet self-defense concepts may, however, obscure rather than clarify that modification.

The Soviets have contended in the UN Outer Space Committee that the Outer Space Treaty requires an agreement between sending and receiving states before direct satellite transmission can begin. They fully regard DTB as a space activity and maintain that the terms of Article 9 of the treaty allow for such discussions.[57] To date, the United States has agreed to enter into consultations on the matter should another state request them but has not said that it would halt a broadcast should such consultations fail to produce an agreement. To do otherwise would, according to American legal experts, constitute prior restraint in violation of the First Amendment.[58] In the meantime, the United Nations has not moved closer to a resolution of the problem.

Although Soviet fear of DTB is genuine, the likelihood of destructive Soviet countermeasures against "unauthorized" broadcasts seems remote. Signal overspill can be reduced or prevented by separating widely the broadcast frequencies used by neighboring states, and receivers can be manufactured or adapted to receive domestic frequencies only. It is hard to imagine an "unauthorized" broadcast requiring more forceful Soviet measures, although a program containing "war propaganda" beamed at Eastern Europe or at Soviet minority nationalities could, if combined with actual troop movements and the like, rise to the level of aggression in Soviet eyes.[59]

Orbital Rendezvous: Rules of Access and Approach

As spacecraft maneuverability has increased and near-earth orbits have become more congested, the question of rules governing the approach of one space object to another has received growing Soviet attention. The importance of developing legal regulations for close approaches derives in part from the difficulty of distinguishing—from the ground or from space—the specific intentions of an approaching spacecraft. For example, problems with one satellite's orbit may bring that object onto a near collision course with another satellite. Or, one country may maneuver a satellite close to the space object of a foreign state for surveillance purposes, thereby raising the alarm of that state. In attempting to develop a legal framework for these situations, as well as similar future ones on the moon, the Soviets have worked out a number of rules of access and visitation [*pravo dostupa i pravo poseshchenia*].

In the last few years, the Soviets have expressed concern about the harm that could result to spacecraft from excessively close, or unplanned, encounters. A leading Soviet specialist has described how "even short-term stationing. . .in the vicinity of a satellite, which, as a rule is equipped with rather sensitive scientific devices. . .may result in interference, and substantially affect satellite performance."[60] The same writer emphasizes the need to devise some scheme of regulation so as to prevent "the involuntary or pre-planned closing in on spacecraft belonging to another State, attempts at interfering with their function, direct damage, or even seizure of the spacecraft in question." Such acts would certainly trigger the right to self-defense.[61]

The Soviet Union contends that notions of state sovereignty in international law bar the "inspection" or close approach of a space vehicle without the consent of the parent country. A space law specialist argues that this prohibition even applies to approaches for purposes of monitoring a present or future arms control agreement. There is no need for such maneuvers, he contends, "insofar as modern technology allows a determination through remote means of whether violations of the concluded agreements have occurred. . .without creating a threat to the safety of space flights."[62] Any problems of concealment [*maskirovka*] in space can be taken up with the SCC, as is done with concealment problems generally.

To protect their space objects but still allow for the possibility of authorized visits or approaches, the Soviets have proposed the establishment of 'security zones' [*zony bezopastnosti*] at some distance around their spaceships. Penetration of a zone would require explicit permission from the state with jurisdiction over the "enclosed" spacecraft. Granting access would depend on such factors as the reason for the visit, the technical condition of the "visited" craft, and the safety of the crew, if any.[63] Ideally, a bilateral or multilateral treaty would establish regulations forbidding not only "the unsanctioned visiting of orbital ships and stations, but also maneuvers of flying near, approach, inspection, etc." in the zones.[64]

On the whole, the Soviets' recommendations seem sensible but on closer inspection are of limited usefulness. Although they bear a rough resemblance to Western efforts to create "rules of the road" for spacecraft, Soviet safety zones seem exclusively oriented toward warding off visits by relatively slow but highly maneuverable spacecraft. Indeed, the suspicion remains that the sudden spate of articles on the topic corresponds to recent exaggerated Soviet fears of the American space shuttle. But while the Soviets appear to disregard the ease with which spacecraft could defend themselves against the shuttle,[65] they also appear to ignore the very real threat posed by direct-ascent or co-orbital ASATs, as well as by wayward satellites. In those cases, the threatening space object either is, or is not, on a collision course; questions of intent matter little if the goal is simply to divert the object. Time, rather than the spatial limits advanced by the Soviets, is therefore the crucial factor, since a state can demand consultations if the former is available and can only attempt self-defense if it is not. Limits on speed of approach in rendezvous maneuvers would accordingly prove a more useful constraint.

More helpful in reducing the risk of an antagonistic encounter in space are a number of simpler measures never discussed by the Soviets. First, the Soviet Union has been particularly unwilling to provide meaningful pre- or even post-flight data about the nature of its spaceflights. Second, the Soviet Union has never, even in theory, made a distinction between "dedicated" and "non-dedicated" ASAT systems so as to facilitate first-stage negotiations on an ASAT ban. Finally, if Soviet conduct surrounding the crashes of radar reconnaissance satellites Kosmos 956 and 1402 is any indication, Soviet willingness to initiate and participate actively in crisis consultations appear somewhat in doubt.

The Space Shuttle

Whatever their actual fears about the American direct-ascent ASAT, the Soviets have devoted much more effort to discussing and criticizing the space shuttle [*kosmicheskii chelnok*]. It may be that an enemy ASAT, no matter how superior its technology, simply does not represent a threat to a country whose military dependence on space will remain relatively smaller for the foreseeable future. The shuttle, on the other hand, with its striking maneuverability and versatility, raises the specter of all kinds of meddling in Soviet space activities and the disclosure of information—including technical shortcomings—that the Soviets have zealously tried to keep secret over the years.

Recently, both the Soviet foreign policy press and legal publications have become quite exercised about the shuttle. The former has focused in some detail on the ability of the shuttle to serve as a platform for weapons and to ferry them between orbital stations.[66] Citing fears about the shuttle inspecting and tampering with Soviet satellites, the journal *Mirovaia ekonomika i mezhdunarodnie otnosheniia* (World Economics and International Relations) also denounced the United States' "taking on the functions of a self-appointed 'space gendarme.'"[67] Soviet lawyers, meanwhile, have in sober tones recommended the negotiation of some kind of international agreement, including the creation of "space traffic rules" [*sootvetstvuiushchikh pravil kosmicheskikh peredvizhenii*] to order the "uncontrolled maneuvering" of reusable space transport systems [*mnogorazovye transportnye kosmicheskie korabli,* or *MTKK*].[68]

For the most part, specific Soviet criticisms of the shuttle and related recommendations have been rendered irrelevant or superfluous with the passage of time. While the Soviets had once attacked shuttle flights as violating the sovereign airspace of foreign countries due to the shuttle's potential horizontal launch as a winged two-stage aircraft, the subsequent use of a vertical lift-off mooted that problem.[69] As for the Soviets' call for a "special international accord with respect to forbidding the placement of weapons of mass destruction on space entities of the *MTKK* type," that objective would already seem to have been fulfilled by the Outer Space Treaty and SALT II.[70]

Indeed, only the arming of a shuttle or its use as a transport of arms into space poses any potential danger to the Soviet Union. That eventuality, particularly in an era of laser and particle beam weapons, could indeed provoke an arms race and make space a very dangerous arena. Although a prohibition on armed shuttles would doubtless fall within the terms of any comprehensive ASAT agreement, the Soviets have specifically urged the adoption of an international understanding "forbidding the location on *MTKK* not only of types of mass destruction weapons, but also any other types of weapons."[71] Although the Soviet position might change upon future acquisition of an operational shuttle, the Soviet Union seems content to criticize the American prototype for the time being.

It is hard to assess the current Soviet view of the shuttle. At one point the Soviets referred to the American shuttle as an ASAT, presumably focusing on its ability to rendezvous and potentially interfere with Soviet satellites.[72] Today it is unclear precisely how they classify it. Certainly no Soviet publication or spokesperson has explicitly acknowledged that it is *not* an ASAT. The 1981 Soviet Draft Treaty banning weapons from space seems to distinguish between "weapons" and vehicles that could be used to carry them. As with satellite reconnaissance, however, a certain amount of negative propaganda can be expected to accompany any Soviet references to technology it has yet to acquire.

THE SOVIET DRAFT SPACE TREATIES

On 11 August 1981, the Soviet Union tabled at the United Nations a Draft Treaty on the Prohibition of the Stationing of Weapons of Any Kind in Outer Space. The document represented the first major statement by one of the superpowers since the Soviet–American ASAT talks had broken off in 1979. The Soviets had come to those talks, despite their ASAT advantage, to listen with curiosity to what the Americans had to say. During the discussions, the parties dealt with several preliminary questions but did not reach any substantive agreement. The Soviets, who reportedly were surprisingly unprepared on a number of topics, were willing to discuss definitions and "rules of the road" but refused to acknowledge ASATs as dangerous for the future uses of space. They similarly balked at a suggestion that they eliminate current asymmetries by dismantling their existing ASAT-equipped rockets. Indeed, the Soviets never admitted to having any ASAT capability in the first place.[73]

In the meantime, three major Soviet concerns also proved to be stumbling blocks. First, the Soviets branded the shuttle an ASAT and insisted on an end to shuttle testing as a quid pro quo for any moratorium on ASAT development. Second, the Soviets sought protection only for satellites of the two parties, possibly reflecting Soviet fears of an emerging Chinese satellite capability. The United States, however, wished to include all satellites in which a party "had an interest," thus extending coverage to NATO and other allied spacecraft. Finally, the Soviet Union apparently wanted to exempt from protection satellites engaging in "hostile" action inimical to national sovereignty.[74]

The Soviets' draft treaty, unexpectedly tabled two years later, raised suspicions that they were either careless or not serious about negotiations. Article 1, paragraph 1, declared that:

> States Parties undertake not to place in orbit around the earth objects carrying weapons of any kind, install such weapons on celestial bodies, or station such weapons in outer space in any other manner, including on reusable manned space vehicles of an existing type or of other types which States Parties may develop in the future.

Article 3 held that each party to the treaty would undertake "not to destroy, damage, disturb the normal functioning, or change the flight trajectory of space objects of other States Parties, if such objects were placed in orbit in strict accordance" with Article 1, paragraph 1. Finally, Article 4 allowed parties to verify the provisions of the treaty using national technical means and enjoined them from "placing obstacles" in the way of such monitoring.[75]

The provisions of Article 1 posed several problems. First, the proposal failed to require the destruction of current ASAT systems, leaving open their possible future use. That came as no surprise, since even if the Soviets wished to dismantle their ASAT system, Western verification of the same might require intrusive on-site inspections. Similarly, the treaty did not ban the testing and deployment of ASATs or other weapon systems, permitting their improvement and a continuing arms race. Also, despite professed Soviet concerns about the shuttle and the treaty reference to it, the proposal might not have barred a quick suborbital sortie by an armed vehicle.[76]

Indeed, Article 1's language about "placing in orbit" or "stationing" weapons in space would not have covered the American direct-ascent F-15/MV system or conceivably even the Soviet interceptor, used in a fractional orbit or "pop up" mode (see Chapters 3 and 4 of this volume). Finally, Article 1 placed no restrictions on other potential ground-based systems, including lasers.

Perhaps the most troublesome aspect of the 1981 Draft Treaty concerned the combined operation of Articles 1 and 3. Article 3 provided that no party may take any hostile action toward space objects "*if* such objects were placed in orbit in strict accordance" with Article 1, paragraph 1 (emphasis added). Viewed as a whole, such conditional phrasing tends to act as a general limitation on satellites rather than as an arms control device against ASATs. In fact, Article 3 seemed to legitimize attacks against satellites engaged in certain acts, rather than forbid such attacks. Such a construction could be used to justify tampering with other countries' space objects on the grounds that such vehicles were suspected to carry weapons.

The USSR's apparent nonchalance in advancing its 1981 draft[77] gave way to a sense of urgency in mid-August 1983. During a visit of American senators to the Soviet Union, Yuri Andropov proposed a moratorium on the deployment of space weapons, pledging that the Soviet Union would unilaterally refrain from such deployments so long as other nations followed suit. Equally significant was Foreign Minister Andrei Gromyko's submission to the United Nations General Assembly of a new, more sophisticated Soviet space arms control treaty.

Titled a "Draft Treaty on Banning the Use of Force in Space and From Space With Respect to the Earth," the new Soviet proposal addresses many of the defects in the previous draft treaty. Article 1 is a broad prohibition on the use of force "with regard to space objects orbiting the Earth, stationed on celestial bodies, or deployed in space in any other manner." Article 2 makes this provision concrete by calling for a comprehensive ban on the testing, deployment, and use of "space objects orbiting the earth, stationed on celestial bodies, or *deployed in space in any other manner* as a means for hitting any targets on the Earth, in the atmosphere, and in space." (Emphasis added.)

Unlike the predecessor document, however, no conditions are attached to a flat pledge "not to destroy, damage, or disrupt the normal functioning of other states' space objects, nor change their flight trajectories." The treaty further demands that signatories not test or develop new antisatellite systems and that they "eliminate such systems already in their possession."[78] One of the more intriguing features of the proposal is that the Soviet Union has for the first time used the term "antisatellite" in an official document and by implication has admitted in the Soviet press that it has an existing ASAT system.[79]

The Soviet treaty contains only a few provisions which the United States and other Western nations should find objectionable. First, it is uncertain how either the United States or the Soviet Union could verify the dismantling of current ASAT systems.[80] Second, the new Soviet draft treaty contains a provision compelling signatories "not to test or use for military, including antisatellite, purposes, any manned spacecraft." This stipulation, recalling as it does possible East–West misunderstandings about "peaceful" and "military" purposes, could unduly restrict use of the shuttle and other manned craft for defensive, nonaggressive missions. Third, Article 6 of the new proposal curiously allows parties to the treaty to "adopt any internal measures which it considers necessary. . . to prohibit and prevent any activity contravening the provisions of this treaty which comes under its jurisdiction or control, wherever it may be." Although it speaks of "internal measures," Article 6 could provide a loophole for all kinds of broad self-defense actions justified on the grounds of state sovereignty.

Ironically, the biggest obstacle to Western acceptance of the new Soviet proposals may well be their comprehensiveness. Although a ban on testing and "hostile uses" would prove immediately verifiable for the most part, a ban on possession and/or deployment would not. At the same time, the treaty provisions that implicate space-based ballistic missile defense and other potential laser weapon applications may meet resistance from those in the United States who favor further development of such systems or who wish to have separate negotiations—and additional bargaining leverage—on that subject.[81]

In weighing the Soviet proposal, American policymakers need to decide, first, whether they want any kind of ASAT agreement and, if that decision is affirmative, whether they want the kind of agreement that Moscow has proposed. There is also the matter of negotiating forum. A bilateral agreement may be called for, since only the two superpowers have ASAT programs at this time, and unavoidable delays would attend negotiation in multilateral forums. Although it could later become the basis for a multilateral treaty, a bilateral framework also allows the countries to build on the success of the SALT Standing Consultative Commission as a vehicle for implementing a space arms control agreement. The provisions of a bilateral ASAT accord, however, should not be structured so as to appear to sanction attacks upon the space objects of third nations. Because the Soviets so clearly regard treaty terms as establishing the only legal obligations of the state, it behooves Western negotiators to be as precise as possible in defining such provisions.

CONCLUSIONS

Soviet space law is an important "verbal policy" in the ongoing East–West ideological struggle. The West can choose to engage the struggle at this level but ought to pick its fights with some attention both to Soviet security concerns and to the relative costs and benefits. For example, the United States stands to gain nothing from insisting on "freedom of information" in the context of DTB except the escalation of future crises. On the other hand, the United States might reap security benefits of its own from agreed procedures to reduce the risk of confrontation in space. Greater information sharing about space military activities and arrangements analogous to Soviet "space safety zones" and to the 1972 Incidents at Sea Agreement could lend more order to the increasing congestion in outer space.[82]

Surveying the whole of Soviet space law relating to military activities, it becomes clear that Soviet policy has a pronounced defensive character. Although the Soviets have manipulated general or ambiguous legal terms to their "offensive" advantage, the bottom line of Soviet legal and political efforts has been controlling the free activity of the West in space. Despite the Soviets' pioneering role in space, Soviet space policy reflects an ongoing battle to rein in threatening American technological superiority—from satellite reconnaissance, to DTB, to the maneuverability of the space shuttle.

Notwithstanding the occasionally ominous tone of Soviet statements on Western use of outer space, the Soviets have acted with restraint during more than two decades of human space exploration. The historical record suggests that the Soviet Union has sought to restrict Western military uses of space only until it could acquire comparable capabilities. Thus, Soviet representatives complained loudly and menacingly about satellite reconnaissance in the early 1960s but grew silent on the subject by the end of the decade as Soviet surveillance from space improved. So, too, it is likely that references to reusable space vehicles have been dropped from the latest Soviet draft space treaty in anticipation of their own space shuttle.

Soviet strategy toward the West in space can be summed up by the phrase "no unilateral advantage." This policy appears to comprehend not only a determination to achieve parity with the West in space—for purposes of pride as well as security—but also a desire to take up any perceived or actual disparities in the SCC, where they perceive superpower parity to have been institutionally enshrined. The Soviets have recommended that the United States and the Soviet Union specifically deal with questions of satellite reconnaissance in that forum and have suggested that both countries utilize the SCC in monitoring compliance with their latest draft treaty.[83]

The rare initiative shown by the Soviet Union in tabling the two draft space treaties may indicate that parity, at least in Soviet eyes, may soon be lost through new technology. Although they have not referred to it often, the Soviets clearly do not want the United States to deploy the MHV. Even more do they fear a space-based ballistic missile defense, which appears to threaten parity on earth as well as in space. To achieve an ASAT test ban, the Soviets seem willing to give up their current ASAT capability.

Due, perhaps, to their lesser reliance on space support systems for the management of military forces, the Soviets never have been very clear about the role of outer space in national security. On the other hand, as evidenced by their ambitious space program and frequent references to a permanent human presence on the moon, and on space stations, the Soviets have much to lose from the deployment of space weapons. If the Soviet Union maintains this prestigious agenda for space, Soviet incentives for space arms control could prove strong in years to come.

NOTES

1. Oleg Khlestov is only one of several Soviet international lawyers who have played a prominent role in conducting aspects of Soviet foreign policy. For example, Vladimir Semyonov, long a prominent member of the Institute of State and Law, served as chief negotiator at the SALT I talks and at the beginning of the SALT II negotiations. He

then became Ambassador to Germany. Stepan Molodtsov, a prominent international legal scholar and Senior Scientific Officer at the Legal Institute of the USSR Academy of Sciences, has played a prominent role at the UN Law of the Sea Conferences. Anatolii Movchan, First Secretary of the Ministry of Foreign Affairs, is a lawyer who has led the Soviet counterattack on human rights. Grigorii Tunkin, the dean of Soviet international lawyers, has held important positions at a number of multinational conventions and at the United Nations. Although it would be a mistake to exaggerate the importance of Soviet international legal experts in foreign policy decisionmaking relative to other groups such as the military, jurists nevertheless have exerted significant policymaking influence in the domestic arena. See, for example, D. Glagolev, "The Soviet Decision-Making Process in Arms Control Negotiations," *Orbis* 21 (1978): 769; Thomas Wolfe, *The SALT Experience* (Cambridge, Mass.: Ballinger, 1979), pp. 60–61; M. McCain, "Soviet Lawyers in the Reform Debate: Cohesion and Efficacy," *Soviet Studies,* 34 (1982): 3–31; D. Barry, "The Specialist in Soviet Policy Making: The Adoption of a Law," *Soviet Studies* 16 (1964): 152; and D. Barry and H. Berman, "The Jurists," in H. Skilling and F. Griffiths, eds., *Interest Groups in Soviet Politics* (Princeton: Princeton University Press, 1971), pp. 122–46.

As for international legal specialists, many of whom are associated with the Institute of State and Law in the USSR Academy of Sciences, their influence in drafting international agreements and participating in international organizations has appeared to grow as the Soviet Union has placed increasing reliance generally on *mezhdunarodniki,* or foreign policy specialists. See Wolfe, pp. 66–68; and Morton Schwartz, *Soviet Perceptions of the United States* (Berkeley: Univ. of California Press, 1978), p. 162–66. Despite the possibility of some compartmentalization of Soviet international law specialists within the foreign policy establishment, the uniformity of Soviet publishing practices and the avowedly political mission of Soviet international law ensures that the legal views expressed reflect official Soviet policy.

2. A. Nikolaev, "Soviet Foreign Policy: Basic Ideological Principles," *International Affairs* (Moscow) 11 (November 1973): 63; also Bernard Ramundo, *Peaceful Coexistence: International Law in the Building of Communism* (Baltimore: The Johns Hopkins University Press, 1967), pp. 22–28. (Hereinafter cited as *Peaceful Coexistence.*)
3. G. Tunkin, "Introductory Article," 1959 *Sovetskii ezhegodnik mezhdunarodnogo prava* (Soviet Yearbook of International Law; hereinafter cited as *Ezhegodnik*) (Moscow: Akademiia nauka, 1960), pp. 9, 13.
4. *Peaceful Coexistence,* p. 9.

5. The provocative/ordering dichotomy is borrowed from Richard Erickson, *International Law and the Revolutionary State* (Dobbs Ferry, New York: Oceana Publishers, 1972, pp. 161–62.
6. For a more detailed examination of this phenomenon, see Mitchell and Leonhard, "Changing Soviet Attitudes Toward International Law: An Incorporative Approach," *Georgia Journal of International Law* 6 (1976): 227.
7. The five major principles of Peaceful Coexistence are (1) Denunciation of war as a method of dispute resolution; (2) Equality of states, mutual trust and understanding between states, and consideration of mutual interests; (3) Nonintervention in the internal affairs of other countries and the right of each country to decide, independently of others, its own internal matters; (4) Respect for the territorial integrity and sovereignty of all states; and (5) The development of economic and cultural cooperation on the basis of full equality and mutual advantage. Kazimierz Grzybowski, *Soviet Public International Law* (Leyden: A.W. Sijthoff, 1970), p. 30. See also *Peaceful Coexistence,* pp. 29–30.
8. Soviet reluctance to subscribe fully to certain traditional principles of international law has made Western negotiators predictably wary of concluding agreements with the Soviet Union. Through experience, they have learned that the "spirit of an agreement" has little, if any, meaning for the Soviets. Now matter how much stock the Soviet Union puts in its own broad statements about international legal standards, only the specific terms of an agreement matter. Unilateral statements and interpretations have little binding effect and cannot be relied on to influence Soviet behavior. *Briefing on SALT I Compliance: Hearing Before the Senate Committee on Foreign Relations,* 96th Cong., 1st sess. (1979), pp. 13–14. Statement by Ambassador Sidney Graybeal, former U.S. Commissioner, SALT Standing Consultative Commission.

   Too often, general phrases may give rise to divergent interpretations, that is, "outwardly synonymous terms can be and are used both in progressive and reactionary sense." M.I. Lazarev, "International Law Terminology and Influence Thereon of the October Revolution," 1968 *Ezhegodnik* (Moscow, 1969): 158–59. A treaty with the Soviet Union will only prove effective if truly reciprocal benefits and incentives for compliance exist on each side. For example, the Helsinki accords have been fraught with difficulties for the West because the Soviet Union has no specific incentives to comply with many key provisions. SALT I and the Nuclear Non-Proliferation Treaty, on the other hand, have generated reasonably good compliance by both the United States and the Soviet Union, dictated by the basic self-interest of both nations. J. Hazard, "Soviet Tactics in International Law-making," *Denver International Law Journal* 7 (1977): 9, 11-13.

9. Unwritten or customary law is still distrusted as a capitalist creation, whereas treaties allow the Soviets to take an active part in the negotiation of an entire range of issues and to adopt tough bargaining positions. The possible inclusion of "peaceful coexistence" or similar terminology in the text or preamble of a treaty also enables the Soviet Union to gain international ratification of important socialist legal principles. By their own admission, the Soviets have concluded as many treaties as possible and today continue to express this preference for treaty negotiation or discussion by frequently referring to the need for "businesslike talks."
10. It is unclear to what degree this characteristic is attributable to the tendency of continental civil law countries to codify as much as possible, or to Soviet defensive tactics, making certain the legal limits of other states' activities. See also J. Triska and R. Slusser, *The Theory, Law and Policy of Soviet Treaties* (Stanford: Stanford University Press, 1962); and Erickson, pp. 72–80.
11. *Briefing on SALT I Compliance,* p. 31 (statement by Amb. Graybeal) and p. 40 (statement by Ambassador Robert Buchheim, also a former U.S. Commissioner to the SCC).
12. *Mezhdunarodnoe kosmicheskoe pravo* (International Space Law) A. Piradov, ed. (Moscow: Mezhdunarodnie otnosheniia, 1974), p. 14 (hereinafter cited as *Kosmicheskoe pravo*); and Y. Korovin, "International Status of Cosmic Space," *International Affairs* (Moscow) 1 (1959): 58 (hereinafter cited as "International Status").
13. V. Vereshchetin, "On the Principle of State Sovereignty in International Space Law," *Annals of Air and Space Law* (Montreal: Carswell Co., 1977), p. 429 (hereinafter cited as "Sovereignty").
14. *Kosmicheskoe pravo,* p. 9.
15. G. Zhukov, "Practical Problems of Space Law," *International Affairs* (Moscow) 5 (May 1963): 28–29; and Stephen Gorove, *Studies in Space Law: Its Challenges and Prospects* (Leyden: A.W. Sijthoff, 1977), pp. 86, 90 (hereinafter cited as *Studies*). It should, however, be noted that the Russian word for military, *voennoe,* carries a far more warlike meaning than its English counterpart. Consequently in Russian the term "nonmilitary" tends to capture more viscerally the idea of "peaceful" than does the word "nonaggressive" (*neagressivnoe*), which is awkward, seldom used, and borrowed from the West.
16. See Chapter 3 for a full discussion of the Soviet military space program.
17. In fact, the Soviets first proposed GCD in 1958. See Y. Korovin, "On the Neutralization and Demilitarization of Outer Space," *International*

Affairs (Moscow) 12 (1969): 82. For an overview of many of the Soviet proposals, see Allan Gotlieb, *Disarmament and International Law* (Toronto: Canadian Institute of International Affairs, 1965); and A. Nikolaev, "International Cooperation for the Peaceful Use of Outer Space," *International Affairs* (Moscow) 5 (May 1960): 76, 80.

18. In the middle of the 1961 Berlin Crisis, Khrushchev boasted that "we placed Gagarin and Titov in space and we can replace them with bombs that can be directed to any place on earth." Quoted in L. Freedman, "The Soviet Union and 'Anti-Space Defense,'" *Survival* 19 (Jan.–Feb. 1977): 16.
19. Feliks Kovalev and Ivan Cheprov, *Na puti k kosmicheskomu pravu* (The Way to Space Law) (Moscow: Institut Mezhdunarodnix otnoshenii, 1962), p. 81 (hereinafter cited as *Na puti*).
20. V.S. Vereshchetin, "Basic Principles of Space Law," 1966–67 *Ezhegodnik* (Moscow, 1968), pp. 125; and *Kosmicheskoe pravo,* pp. 78–79.
21. Inexplicably, this description occurs only in the English edition of the standard Soviet treatise on space law. *International Space Law,* A.S. Piradov, ed. (Moscow: Progress Publishers, 1976), p. 94. For a similar early formulation see also G. Zhukov, "The Demilitarization and Neutralization of Outer Space," *Sovetskoe gosudarstvo i pravo* (Soviet Government and Law) 5 (1962): 265.
22. V.S. Vereshchetin, "Against Arbitrary Interpretation of Some Important Provisions of International Space Law," *Sovetskoe gosudarstvo i pravo* 5 (1983): 81 (hereinafter cited as "Arbitrary Interpretation").
23. *Kosmicheskoe pravo,* p. 121.
24. A.I. Rudev, "Preventing the Militarization of Space," *Sovetskoe gosudarstvo i pravo* 1 (1983): 66 (hereinafter cited as "Militarization").
25. See Evan Luard, *The Control of the Seabed* (New York: Taplinger Publishing Co., 1974), pp. 97–112; also, T. Treves, "Military Installations, Structures, and Devices on the Seabed," *American Journal of International Law* 74 (1980): 808.
26. But see footnote 79, *infra.*
27. TASS, 4 April 1977, Joint Publications Research Service (hereinafter cited as JPRS), *USSR Report: Space*, No. 56907 (14 June 1977).
28. V. Rostarchuk, "The Shuttle in the Pentagon's Plans," *Krasnaia zvezda* (29 July 1981): 5.
29. Soviet foreign minister Maxim Litvinov first urged the "priority" formulation on the world community in the 1930s, a telling reflection, according to some, of the traditional Russian inability to mobilize quickly in advance of foreign attack. Edward McWhinney, *The International Law of Detente* (Alphen aan den Rijn: Sijthoff and Noordhoff, 1978), p. 175. It may also reflect enduring Soviet fears about "capitalist encirclement."

30. McWhinney, p. 176. The prevailing Western view stems from the American experience with the Cuban Missile Crisis, where the United States wished to take preemptive actions against preparations for an enemy strike without such action being automatically viewed as agression. C. Oliver, "International Law and the Quarantine of Cuba: A Hopeful Presumption for Legal Writing," *American Journal of International Law* 57 (1963): 375.
31. U.N. General Assembly Resolution 3324 (XXIX), 29 GAOR, Suppl. 142, UN Doc. A/9631 (1975).
32. See, for example, M. McDougal and F. Feliciano, *Law and Minimum World Public Order—The Legal Regulation of International Coercion* (New Haven: Yale University Press, 1961), pp. 231–41.
33. V. Vereshchetin, "Arbitrary Interpretation," p. 83.
34. Y. Kolosov, *Borba za mirnyi kosmos* (Struggle for a Peaceful Outer Space) (Moscow: Mezhdunarodnie otnosheniia, 1968), p. 49 (hereinafter cited as *Borba*).
35. V. Sokolovskii, *Soviet Military Strategy,* 3rd edition. Edited and with an analysis and commentary by H.F. Scott. (New York: Crane, Russak and Co., 1975), pp. 179–89, 280; and S. Meyer, "Anti-Satellite Weapons and Arms Control: Incentives and Disincentives from the Soviet and American Perspective," *International Journal* 36 (1981): 66–67.
36. G.P. Zhukov, "Space Espionage Plans and International Law," *International Affairs* (Moscow) 10 (1960): 56. For more on the satellite/aircraft analogy, *see also* F. Kovalev and I. Cheprov, "Concerning the Elaboration of Legal Problems of Outer Space," *Sovetskoe gosudarstvo i pravo* 7 (1960): 132–33.
37. G.P. Zadorozhnyi, "The Basic Problems of the Science of Cosmic Law, in *Kosmos i mezhdunarodnoe pravo,* E.A. Korovin, ed. (Moscow: Institut Mezhdunarodnix otnoshenii, 1962), p. 53.
38. G.P. Zhukov, "Nuclear Demilitarization of Outer Space," *Sovetskoe gosudarstvo i pravo* 3 (1964): 83.
39. V. Aniutin, "Surveillance of Outer Space," *Voennaia mysl* 3 (1968).
40. V. Viktorov, "An Agreement of Historic Importance," *Mezhdunarodnaia zhizn* (Moscow) 7 (1972): 25.
41. For an excellent survey of the entire Soviet approach to space reconnaissance in the 1960s and 1970s, see S. Cohen, "SALT Verification: The Evolution of Soviet Views and Their Meaning for the Future," *Orbis* 24 (1980): 657– .
42. Y. Safronov and Y. Sukhanov, "Samos, Midas and Others," *Krasnaia zvezda* (26 October 1972): 3.
43. V. F. Lobanov, "On Guard Over the Fatherland's Frontiers," *Krasnaia zvezda* (27 May 1972): 3.

44. I. Kotliarov, "International Space Monitoring for Peace," (hereinafter cited as "Monitoring"). In M.D. Schwartz, ed., *Proceedings of the* [21st] *Colloquium on the Law of Outer Space.* Sponsored by the International Institute of Space Law of the International Astronautical Federation (hereinafter cited as *Colloquium*). Littleton, Colo.: Fred B. Rothman & Co., 1978), p. 166. Kotliarov has devised a four-point standard by which to judge permissible monitoring. It must be based on (1) a treaty or other agreement confirming state power to engage in monitoring; (2) matching the functions of reconnaissance to the obligations of treaties; (3) ensuring that the extent of monitoring corresponds to the extent of the obligations; and (4) guaranteeing the "simultaneity" of monitoring obligations—that is, ensuring that monitoring by both sides changes in conformity to agreed on modifications or revocations. Idem, pp. 166–67.
45. I. Kotliarov, *Mezhdunarodnyi kontrol c ispolzovaniem kosmicheskikh sredstv* (International Monitoring with the Use of Space Means) (Moscow: Mezhdunarodnie otnosheniia, 1981), p. 95 (hereafter cited as *Kontrol*).
46. The treaty terms accepted by both the United States and the Soviet Union under SALT I appear to permit nationwide surveillance. Since the ABM treaty in Article I prohibits a national ABM defense system—requiring an extensive warning system and communications network—it is hard to imagine an effective verification regime that would not involve satellite reconnaissance over the whole of Soviet territory.
47. See, for example, the confidentiality clauses contained in the memorandum of understanding of the Standing Consultative Commission established under SALT I, U.S. Dept. of State, "Standing Consultative Commission on Arms Limitation: Regulations," *United States Treaties and Other International Agreements* (T.I.A.S. 7637), (1974), pp. 1124–.
48. See V. Bordunov, "Some Legal Problems of Remote Sensing of Earth from Outer Space," 20th *Colloquium* (1977), pp. 496–97; and G.P. Zhukov, "International Law Problems Related to the Exploration of Earth Resources from Outer Space," 19th *Colloquium* (1976), p. 112.
49. Kotliarov, *Kontrol,* p. 96.
50. Kotliarov, "Monitoring," p. 166. It should be noted that despite its willingness to take certain issues to the United Nations, the Soviet Union has firmly opposed the establishment of a multilateral verification body, including the recent French proposal for an International Satellite Monitoring Agency (ISMA). M. Abdel-Hady and A. Sadek, "Verification Using Satellites: Feasibility of an International or Multinational Agency," in B. Jasani, ed., *Outer Space: A New Dimension*

of the Arms Race (Cambridge, Mass.: Oelgeschlager, Gunn and Hain, 1982), pp. 293–94.

51. In 1975, there was concern that U.S. early-warning satellites had been attacked by Soviet infrared lasers; later it was established that these satellites had only observed infrared radiation from a Soviet natural gas pipeline fire. *Briefing on SALT I Compliance,* p. 54.
52. *Kosmicheskoe pravo,* p. 175.
53. Idem, p. 176.
54. G.P. Zhukov, *Kosmicheskoe pravo* (Space Law) (Moscow: Mezhdunarodnie otnosheniia, 1966), p. 216.
55. United Nations (U.N.) Document A/8771, 8 August 1972.
56. U.N. Doc. A/AC.105/PV.127, Annex 2, 2 April 1974, pp. 1–2; Zhukov, "International Law Problems of Direct Television Broadcasting," 19th *Colloquium* (1976), p. 118.
57. Legal Sub-committee of the COPUOS, 29 U.N. GAOR (218th mtg.), U.N. Doc. A/AC.105/C.2/SR. 218 (1974), p. 61; Legal Subcommittee of the COPUOS, 30 U.N. GAOR (239th mtg.), U.N. Doc. A/AC.105/C.2/SR.239 (1975), p. 115.
58. Legal Subcommittee of the COPUOS, 33 U.N. GAOR (288th mtg.), U.N. Doc. A/AC.105/C.2/SR.288 (1978), pp. 5–6; Carl Christol, *The Modern International Law of Outer Space* (New York: Pergamon Press, 1982), pp. 688–89.
59. One Soviet jurist implicitly refers to jamming as a means of dealing with the DTB threat, G.P. Zhukov, "Actual Problems of International Space Law," 1977 *Ezhegodnik* (Moscow, 1978): 132, while the most recent edition of the standard treatise on international law conceivably recognizes a variety of self-defense measures in declaring that the receiving state may "take measures consistent with international law." In F.I. Kozhevnikov, ed., *Mezhdunarodnoe pravo* (International Law) (Moscow: Mezhdunarodnie otnosheniia, 1981), p. 194.
60. B.G. Dudakov, "On the International Legal Status of Artificial Earth Satellites and the Zones Adjacent to Them," *Proceedings of the 24th Colloquium on the Law of Outer Space* (1981) (New York: American Institute of Aeronautics and Astronautics, 1982), p. 100.
61. V.S. Vereshchetin, "On the Elaboration of Legal Problems of Manned Space Flights," 1981 *Ezhegodnik* (Moscow, 1982), p. 169.
62. Rudev, "Militarization," pp. 68–69.
63. A. Rudev, *Mezhdunarodno-pravovoy status kosmicheskikh stantsii* (The International Legal Status of Space Stations) (Moscow: Mezhdunarodnie otnosheniia, 1982), p. 74.
64. A.I. Rudev and P. Lukin, *Kosmos i pravo* (Space and Law) (Moscow, 1980), p. 104.
65. It is obvious that mines or other "booby traps" on a Soviet space object would act as a powerful deterrent to close inspection or seizure by the

prohibitively expensive Shuttle. While the United States has noted that nothing in the Outer Space Treaty specifically precludes close inspection by one state of the space objects of another, several witnesses at the Senate hearings on the treaty acknowledged that national technical means, though inferior, would be relied on to monitor compliance with the agreement. *Treaty on Outer Space: Hearings Before the Senate Committee on Foreign Relations,* Executive D, 90th Cong., 1st sess. (1967), pp. 22, 93 (statements by Mr. Goldberg and Mr. Vance, respectively).

66. V. Basmanov, "For a Weapons-Free Outer Space," *International Affairs* (Moscow) 11 (1981): 102; and S. Stashevskii, "The USSR in the Struggle for a Peaceful Outer Space," *International Affairs* (Moscow) 7 (1981): 62–63.
67. S. Stashevskii & G. Stakh, "Outer Space Must Be Peaceful," *Mirovaia ekonomika i mezhdunarodnie otnosheniia* 2 (1982): 19.
68. A.I. Rudev, "Space Shuttle Program: Political and Legal Problems," *Sovetskoe gosudarstvo i pravo* 4 (1981): 89–90 (hereinafter cited as "Space Shuttle").
69. Because the shuttle could be used somewhat like a conventional aircraft, the Soviets continue to press for regulations covering flights through territorial airspace. (Idem, pp. 88–89.) Ironically, the low reentry angle over land of the proposed Soviet shuttle probably would result in the kind of violation of sovereignty complained of by the Soviets. The third flight test of a Soviet scale-model shuttle ended in the Black Sea, having overflown Western and Eastern Europe on reentry. Thomas O'Toole, "Soviets Orbit Space-Shuttle Prototype," *Washington Post,* 28 December 1983.
70. Article 4(1) prohibits the placing in orbit of "any objects carrying nuclear weapons or any other kinds of weapons of mass destruction" or the stationing of "such weapons in outer space in any other manner." Article 9 of SALT II bars the parties from developing, testing, or deploying "systems for placing into Earth's orbit nuclear weapons or any other kind of weapons of mass destruction, including fractional orbiting missiles." See US Arms Control and Disarmament Agency, *Arms Control and Disarmament Agreements,* 1980 Edition (Washington: USGPO, 1980), pp. 52, 225–26.
71. Rudev, "Militarization," p. 67.
72. "Soviets See Shuttle as Killer Satellite," *Aviation Week & Space Technology* (17 April 1978): 17. The Soviets reportedly brought this view with them to the 1978–79 ASAT talks.
73. Interview with Ambassador Herbert York, 28 April 1983; R. Burt, "U.S. Seeks to Curb 'Killer Satellites,'" *New York Times,* 10 April 1979.
74. R. Burt, "Soviet Said to Ask Space Shuttle Halt," *New York Times,* 1 June 1979; York interview; and D. Goedhuis, "Some Observations on

the Efforts to Prevent a Military Escalation in Outer Space," *Journal of Space Law* 10 (1982): 19.

75. On the other hand, Article 4 did not specifically mandate noninterference with national technical means of verification. "Draft Treaty on the Prohibition of the Stationing of Weapons of any kind in Outer Space," 11 August 1981, U.N. Doc. A/RES/36/97, 15 January 1982, pp. 3–5.
76. Article 1, para. 1 of the Draft Treaty bars the stationing of weapons in space "in any other manner, including on reusable manned space vehicles of an existing type..." The idea that the Soviets intended somehow to impair the Shuttle program was reinforced by the reported omission of the word "on" in the English texts of the Soviet Draft Treaty used by some nations at the United Nations, so that "reusable manned vehicles" appeared to fall within the treaty ban.
77. Foreign Minister Andrei Gromyko made only passing reference to the Soviet Draft Treaty in the General Assembly Plenary sessions of 1981 and 1982. See U.N. Doc. A/36/PV.7 (22 September 1981) and U.N. Doc. A/37/PV.12. The fact that the Soviets acceded to the transmittal of their proposal to the Committee on Disarmament, a body with little expertise in the subject, would seem to confirm a certain absence of urgency on the Soviet side. Both the United States and the Soviet Union bypassed COPUOS, which could have had joint or exclusive jurisdiction.
78. "Draft Treaty on Banning the Use of Force in Space and From Space With Respect to the Earth," *Pravda* (22 August 1983): 4. Reported in U.S. Foreign Broadcast Information Service, *Daily Report—Soviet Union* 163 (22 August 1983): AA2–AA4.
79. On the domestic Russian radio network, a commentator observed the following: "Weinberger, the U.S. Secretary of Defense, made a hurried statement to the effect that the Soviet Union does, supposedly, not wish to create antisatellite systems because it already possesses them. Well, the Soviet proposal, and I repeat here, refers not only to the banning of the development of new systems, but also to the destruction of the existing ones." Viktor Levin, "Commentary," Moscow Domestic Service, 1500 GMT, 19 August 1983, reported in Foreign Broadcast Information Service, *Daily Report—Soviet Union* (22 August 1983): AA6.
80. A ban on the possession of ASATs would, for example, probably entail intrusive on-site inspections for both sides, given the small size of the American MHV and the interchangeability of the Soviet ASAT system with a number of components and facilities shared with other Soviet space missions.
81. Even before President Reagan's "star wars" speech in March 1983, the Soviets displayed far greater concern about BMD than about ASAT

warfare. See, for example, *Verbatim Record of the U.N. Committee on Disarmament,* 184th meeting, 2 September 1982, U.N. Doc. A/CD/PV.184, p. 14 (statement by Mr. Issraelian). Chapters 3 and 9 in this volume discuss aspects of Soviet motivation with respect to ASAT development.

82. The Incidents Agreement is discussed in Chapter 9.
83. See Article 5(2) of the Soviet 1983 draft treaty and Kotliarov, "Monitoring," p. 166.

REFERENCES

Abdel-Hady, M., and A. Sadek. "Verification Using Satellites: Feasibility of an International or Multinational Agency." In Bhupendra Jasani, ed. *Outer Space: A New Dimension of the Arms Race.* Cambridge, Mass.: Oelgeschlager, Gunn and Hain, 1982.

Aniutin, V. "Surveillance of Outer Space." *Voennaia mysl* (Military Thought) 3 (1968).

Barry, D. "The Specialist in Soviet Policy Making: The Adoption of a Law," *Soviet Studies* 16 (1964): 152–165.

———, and H. Berman. "The Jurists." pp. 291–333. In H. Skilling and F. Griffiths, eds. *Interest Groups in Soviet Politics.* Princeton: Princeton University Press, 1971.

Basmanov, V. "For a Weapons-Free Outer Space." *International Affairs* (Moscow) 11 (1981): 100–103.

Bordunov, V. "Some Legal Problems of Remote Sensing of Earth from Outer Space." pp. 496–98. In Mortimer D. Schwartz, ed. *Proceedings of the 20th Colloquium on the Law of Outer Space* (1977). Sponsored by the International Institute of Space Law of the International Astronautical Federation. Littleton, Colo.: Fred B. Rothman & Co., 1978.

Burt, Richard. "Soviet Said to Ask Space Shuttle Halt." *New York Times,* 1 June 1979. p. A6.

———. "U.S. Seeks to Curb 'Killer Satellites.'" *New York Times,* 10 April 1979. pp. A1, A17.

Christol, Carl. *The Modern International Law of Outer Space.* New York: Pergamon Press, 1982.

Cohen, S. "SALT Verification: The Evolution of Soviet Views and Their Meaning for the Future." *Orbis* 24 (1980): 657–683.

"Draft Treaty on Banning the Use of Force in Space and From Space With Respect to the Earth," *Pravda,* 22 August 1983. In U.S. Foreign Broadcast Information Service *Daily Report—Soviet Union* 163 (22 August 1983): AA2–AA4.

Dudakov, B.G. "On the International Legal Status of Artificial Earth Satellites and the Zones Adjacent to Them." pp. 97–101. *Proceedings of the*

24th Colloquium on the Law of Outer Space (1981). Sponsored by the International Institute of Space Law of the International Astronautical Federation. New York: American Institute of Aeronautics & Astronautics, 1982.

Erickson, Richard. *International law and the Revolutionary State.* Dobbs Ferry, New York: Oceana Publishers, 1972.

Freedman, Lawrence. "The Soviet Union and 'Anti-Space Defense.'" *Survival* 19 (Jan.–Feb. 1977): 16–23.

Glagolev, D. "The Soviet Decision-Making Process in Arms Control Negotiations." *Orbis* 21 (1978): 767–776.

Goedhuis, D. "Some Observations on the Efforts to Prevent a Military Escalation in Outer Space." *Journal of Space Law* 10 (1982): 13–30.

Gorove, Stephen. *Studies in Space Law: Its Challenges and Prospects.* Leiden: A.W. Sijthoff, 1977.

Gotlieb, Allan. *Disarmament and International Law.* Toronto: Canadian Institute of International Affairs, 1965.

Grzybowski, Kazimierz. *Soviet Public International Law.* Leiden: A.W. Sijthoff, 1970.

Hazard, J. "Soviet Tactics in International Lawmaking." *Denver International Law Journal* 7 (1977): 9–32.

Kolosov, Y. *Borba za mirnyi kosmos* (Struggle for a Peaceful Outer Space) Moscow: Mezhdunarodnie otnosheniia, 1968.

Korovin, Y. "International Status of Cosmic Space." *International Affairs* (Moscow) 1 (1959): 53–59.

———. "On the Neutralization and Demilitarization of Outer Space." *International Affairs* (Moscow) 12 (1969): 82–83.

Kotliarov, I. "International Space Monitoring for Peace." pp. 165–68. In Mortimer D. Schwartz, ed. *Proceedings of the 21st Colloquium on the Law of Outer Space* (1979). Sponsored by the International Institute of Space Law of the International Astronautical Federation. New York: American Institute of Aeronautics and Astronautics, 1980.

———. *Mezhdunarodnyi kontrol c ispolzovaniem kosmicheskikh sredstv.* (International Monitoring with the Use of Space Means) Moscow: Mezhdunarodnie otnosheniia, 1981.

Kovalev, F., and I. Cheprov. "Concerning the Elaboration of Legal Problems of Outer Space." *Sovetskoe gosudarstvo i pravo* 7 (1960): 130–38.

———. *Na puti k kosmicheskomu pravu.* (The Way to Space Law.) Moscow: Institut Mezhdunarodnix otnoshenii, 1962.

Kozhevnikov, F.I., ed. *Mezhdunarodnoe pravo.* (International Law) Moscow: Mezhdunarodnie otnosheniia, 1981.

Lazarev, M.I. "International Law Terminology and Influence Thereon of the October Revolution." pp. 132–58. 1968 *Sovetskii ezhegodnik mezhdunarodnogo prava* (Soviet Yearbook of International Law) Moscow: Akademiia nauka, 1969.

Levin, Viktor. "Commentary." Moscow Domestic Service, 1500 GMT, 19 August 1983. In Foreign Broadcast Information Service, *Daily Report—Soviet Union* 163 (22 August 1983): AA5–AA6.

Lobanov, V.F. "On Guard Over the Fatherland's Frontiers." *Krasnaia zvezda* (Red Star), 27 May 1972. p. 3.

Luard, Evan. *The Control of the Seabed.* New York: Taplinger Publishing Co., 1974.

McCain, Morris. "Soviet Lawyers in the Reform Debate: Cohesion and Efficacy." *Soviet Studies* 34 (1982): 3–22.

McDougal, M., and F. Feliciano. *Law and Minimum World Public Order—The Legal Regulation of International Coercion.* New Haven: Yale University Press, 1961.

Meyer, S. "Anti-Satellite Weapons and Arms Control: Incentives and Disincentives from the Soviet and American Perspective." *International Journal* 36 (1981): 460–84.

McWhinney, Edward. *The International Law of Detente.* Alphen aan den Rijn: Sijthoff and Noordhoff, 1978.

Mitchell, R.J., and A.J. Leonhard. "Changing Soviet Attitudes Toward International Law: An Incorporative Approach." *Georgia Journal of International Law* 6 (1976): 227–44.

Nikolaev, A. "International Cooperation for the Peaceful Use of Outer Space." *International Affairs* (Moscow) 5 (1960): 76–80.

———. "Soviet Foreign Policy: Basic Ideological Principles," *International Affairs* (Moscow) 11 (November 1973): 63–70.

Oliver, C. "International Law and the Quarantine of Cuba: A Hopeful Presumption for Legal Writing." *American Journal of International Law* 57 (1963): 373–77.

O'Toole, Thomas. "Soviets Orbit Space-Shuttle Prototype." *Washington Post,* 28 December 1983. p. A12.

Piradov, A.S., ed. *International Space Law.* Moscow: Progess Publ., 1976.

———. *Mezhdunarodnoe kosmicheskoe pravo* (International Space Law) Moscow: Mezhdunarodnie otnosheniia, 1974.

Ramundo, Bernard. *Peaceful Coexistence: International Law in the Building of Communism.* Baltimore: The Johns Hopkins University Press, 1967.

Rostarchuk, V. "The Shuttle in the Pentagon's Plans." *Krasnaia zvezda,* 29 July 1981. p. 3.

Rudev, A.I. *Mezhdunarodno-pravovoy status kosmicheskikh stantsii* (The International Legal Status of Space Stations) Moscow: Mezhdunarodnie otnosheniia, 1982).

———. "Preventing the Militarization of Space." *Sovetskoe gosudarstvo i pravo* 1 (1983): 62–70.

———. "Space Shuttle Program: Political and Legal Problems." *Sovetskoe gosudarstvo i pravo* 4 (1981): 86–94.

Safronov, Y., and Y. Sukhanov. "Samos, Midas and Others." *Krasnaia zvezda,* 26 October 1972. p. 3.

Schwartz, Morton. *Soviet Perceptions of the United States.* Berkeley: University of California Press, 1978.

Sokolovskii, V. *Soviet Military Strategy.* 3rd edition. Edited and with an analysis and commentary by H.F. Scott. New York: Crane, Russak and Co., 1975.

"Soviets See Shuttle as Killer Satellite." *Aviation Week & Space Technology* (17 April 1978): 17.

Stashevskii, S. "The USSR in the Struggle for a Peaceful Outer Space." *International Affairs* (Moscow) 7 (1981): 62–69.

———, and G. Stakh. "Outer Space Must Be Peaceful." *Mirovaia ekonomika i mezhdunarodnie otnosheniia* 2 (1982): 77–84.

Treves, T. "Military Installations, Structures, and Devices on the Seabed." *American Journal of International Law* 74 (1980): 808–57.

Triska, J., and R. Slusser. *The Theory, Law and Policy of Soviet Treaties.* Stanford: Stanford University Press, 1962.

Tunkin, G. "Introductory Article." pp. 11–15. 1959 *Sovetskii ezhegodnik mezhdunarodnogo prava* (Soviet Yearbook of International Law) Moscow: Akademiia nauka, 1960.

United Nations. Committee on Disarmament. Verbatim Record. 184th meeting. September 2, 1982. A/CD/PV.184. (Statement by Soviet Ambassador Issraelian on space weapons.)

———. Committee on the Peaceful Use of Outer Space. A/AC.105/PV.127, Annex 2, 2 April 1974.

———. Committee on the Peaceful Use of Outer Space. Legal Subcommittee. A/AC.105/C.2/SR.288, 1978.

———. General Assembly. "Consensus Definition of Aggression." G.A. Res. 3314 (XXIX). A/9631 (1975).

———. General Assembly. "Draft International Convention on Principles of Direct Television Broadcasting." A/8771, 8 August 1972.

———. General Assembly. "Draft Treaty on the Prohibition of the Stationing of Weapons of any kind in Outer Space." Submitted by the Soviet Union, 11 August 1981. A/RES/36/97, 15 January 1982.

U.S. Arms Control and Disarmament Agency. *Arms Control and Disarmament Agreements,* 1980 Edition. Washington: Government Printing Office, 1980.

U.S. Congress. Senate. *Briefing on SALT I Compliance: Hearing Before the Committee on Foreign Relations,* 96th Cong., 1st sess. 1979.

U.S. Congress. Senate. *Treaty on Outer Space: Hearings Before the Senate Committee on Foreign Relations,* Executive D, 90th Cong., 1st sess. 1967.

U.S. Department of State. *United States Treaties and Other International Agreements.* "Standing Consultative Commission on Arms Limitation: Regulations." T.I.A.S. 7637 (1974).

U.S. Joint Publications Research Service. *USSR Report: Space,* No. 56907, 14 June 1977. TASS broadcast, 4 April 1977.

Vereshchetin, V.S. "Against Arbitrary Interpretation of Some Important Provisions of International Space Law." *Sovetskoe gosudarstvo i pravo* 5 (1983): 77–84.

———. "Basic Principles of Space Law." 1966–67 *Sovetskii ezhegodnik mezhdunarodnogo prava* Moscow: Akademiia nauka, 1968.

———. "On the Elaboration of Legal Problems of Manned Space Flights." 1981 *Sovetskii ezhegodnik mezhdunarodnogo prava* Moscow: Akademiia nauka, 1982. pp. 166–77.

———. "On the Principle of State Sovereignty in International Space Law." *Annals of Air and Space Law,* Vol. 2. Montreal: Carswell Co., 1977. pp. 429–436.

Viktorov, V. "An Agreement of Historic Importance." *Mezhdunarodnaia zhizn* (Moscow) 7 (1972): 18–29.

Wolfe, Thomas. *The SALT Experience.* Cambridge, Mass.: Ballinger, 1979.

York, Amb. Herbert. Personal interview, 28 April 1983.

Zadorozhnyi, G.P. "The Basic Problems of the Science of Cosmic Law." pp. 23–88. In E.A. Korovin, ed. *Kosmos i mezhdunarodnoe pravo.* Moscow: Institut Mezhdunarodnix otnoshenii, 1962.

Zhukov, G.P. "Actual Problems of International Space Law." 1977 *Sovetskii ezhegodnik mezhdunarodnogo prava.* Moscow: Akademiia nauka, 1978. pp. 186–202.

———. "International Law Problems of Direct Television Broadcasting." pp. 115–18. In Mortimer D. Schwartz, ed. *Proceedings of the 19th Colloquium on the Law of Outer Space* (1976). Sponsored by the International Institute of Space Law of the International Astronautical Federation. Littleton, Colo.: Fred B. Rothman & Co., 1977.

———. "International Law Problems Related to the Exploration of Earth Resources from Outer Space." Ibid. pp. 108–13.

———. *Kosmicheskoe pravo* (Space Law) Moscow: Mezhdunarodnie otnosheniia, 1966.

———. "Nuclear Demilitarization of Outer Space." *Sovetskoe gosudarstvo i pravo* 3 (1964): 79–89.

———. "Practical Problems of Space Law." *International Affairs* (Moscow) 5 (1963): 27–30.

———. "Space Espionage Plans and International Law." *International Affairs* (Moscow) 10 (1960): 53–57.

9 APPROACHES TO THE CONTROL OF ANTISATELLITE WEAPONS

Donald L. Hafner

The desirability of arms control limits on military activities in outer space is not self-evident. The goal of arms control should be to serve American security interests through negotiated restraints. Yet popular enthusiasm for banning weapons from outer space often arises more from general sentiment than from specific security arguments. The focus upon *space* weapons implies that the place where weapons are used, rather than their potential impact upon the probability of war or its ferocity, ought to be the dominant consideration. In popular perceptions, it seems, outer space is a pristine location as yet unsullied by the animosities that plague humanity on earth. The prospect of making space a sanctuary evokes a deep human longing for new beginnings, for transcendence over past corruptions of the human venture.

But outer space became a military arena long ago. Military satellites now routinely scan the globe from pole to pole, and some may sweep as low as 200 kilometers or less above any point on the earth's surface to carry out their missions. In an era of modern weaponry when nations feel compelled to stake their security perimeters hundreds, even thousands of kilometers out from their borders, it would be surprising if outer space could become a sanctuary from traditional security concerns. If a case is to be made for space arms control, it must take these concerns into account.

The prime justifications for putting weapons in space are two: to destroy an adversary's satellites that are engaged in hostile acts and to attack terrestrial targets from space. These are qualitatively different military tasks, and only the first is within the grasp of current technology. Space weapons for attacking targets on earth—such as laser satellites able to destroy ballistic missiles, or even aircraft, ships, tanks, or troops—may rouse the military imagination. But they are still largely speculative, and thus we cannot know how soon, if ever, they will replace more prosaic ways of ensuring our security. Moreover, in some instances these futuristic weapons are already constrained by arms control treaties (see Chapters 6 and 7). On the other hand, superpower competition in weapons for attacking satellites (ASATs) has already begun. These ASAT weapons may do less to fire the imagination, but they constitute the most immediate issue for arms control in outer space.

The manner in which space arms control measures might serve American security interests can best be judged by considering the probable course of U.S. and Soviet innovations in space weaponry over the next decade or so, if no arms control restraints are imposed.

The variety of U.S. military uses of space will certainly expand over the coming decade. Plausible new missions would include radar satellites for all-weather surveillance of land and ocean areas, infrared imaging satellites for detecting and inspecting other satellites in outer space, relay satellites that will transfer data from satellite to satellite and coordinate some satellite missions, and at least rudimentary laser satellites for ASAT or anti-ASAT (DSAT) missions. The total number of U.S. satellites, military and civilian, in service at any time will increase to well over a hundred. If ASAT weapons are otherwise unrestrained, there will be some pressure to place the bulk of military satellites in geosynchronous or highly-elliptical (Molniya-type) orbits, beyond the reach of the most immediate Soviet ASAT threats. Nevertheless, a sizable fraction of satellites—perhaps one-third, mostly military surveillance satellites—will remain in lower orbits (below 3,000 km or so) where their missions can be conducted most effectively.

Innovations in the military use of space will certainly tempt both the United States and the Soviet Union to improve upon their current ASAT systems. The United States will feel itself under somewhat less pressure to innovate because it will begin with a highly-sophisticated miniature ASAT interceptor and a mobile launch platform (F-15 aircraft) with inherent flexibility. Developing a larger ground- or air-launched booster capable of carrying the miniature ASAT to higher altitudes would be a plausible line of improvement. Ground- and space-based lasers will also be pursued, not because they would initially be superior ASATs but because the ASAT mission provides a convenient rationale for funding and testing these exotic technologies.[1]

The Soviets, on the other hand, will be forced to innovate sooner because their current ASAT interceptor is based upon 1960s technology and has limited growth potential. The Soviets could improve their interceptor's homing sensor, add maneuvering capability, multiply the number of launch sites, even strap the interceptor to a larger booster (such as the Proton booster used to launch Salyut) in order to improve its performance, so it could attack more U.S. satellites at higher orbits or defeat U.S. satellite defensive measures. In the end, however, the interceptor would remain an outmoded, cumbersome, and expensive testament to the state of Soviet space technology two decades ago. Since they must innovate, the Soviets may follow accustomed practices and develop a variety of ASAT weapons, each adapted for specific tasks. These could include ground-based lasers for lower orbit attacks, a more agile miniature interceptor such as the U.S. ASAT for direct-ascent attacks on satellites at low and medium orbits, and space "mines" (satellites containing explosives, stationed in orbit during peacetime near their potential targets) for attacks on geosynchronous satellites.

Each improvement in U.S. and Soviet ASAT capability will, in turn, spark efforts at defensive satellite countermeasures, but the competitive edge will almost certainly remain with the offense. Satellites deployed to perform specific missions, and only incidentally equipped to defend themselves, will be at a disadvantage in duels with sophisticated ASATs. Anti-ASAT weapons (DSATs) would offer a more promising line of defense—lasers, interceptors, and space mines that can counterattack lasers, interceptors, and space mines. Such DSAT weapons would work best if they struck preemptively, before an opposing space mine could explode by its target or before an ASAT laser could attack a DSAT laser. In times of confrontation, therefore, close approaches or ambiguous acts by the other side's satellites would raise anxieties about preemption, and the very process of placing ASATs and DSATs on alert as a safeguard against preemption may be indistinguishable from preparations for attack. Even if the United States and the Soviet Union have avoided becoming too dependent upon satellite systems, such unlimbering of ASATs and DSATs may be taken as inflammatory challenges. If dependence upon satellites becomes high, such challenges and the fears they arouse may hasten the pace at which confrontation turns to conflict or conflict to cataclysm.

The fact that such dire consequences *might* follow from unrestrained competition in space weaponry is certainly cause for worrying about future U.S. security. But again it is not a decisive argument for arms control. Even if both superpowers viewed the consequences of arms competition in outer space with foreboding, arms control would work only if limitations could be devised that were effective and verifiable, at an acceptable political price, and did not impair other aspects of security. This may be a difficult standard to meet.

Arms control limits on military activities in space could take many forms. Several formulations will be sketched out in the following pages, ranging from a thorough ban on all military activities, through limits on antisatellite weaponry alone, and ending with mutual pledges simply not to attack the satellites of other states. This is hardly an exhaustive list, but it does sweep in the main features of arms control proposals that have either been put forward by the United States or Soviet Union or seem likely to be in the near future. Moreover, it is a useful list for illustrating the benefits and vexing problems posed by outer space arms control.

COMPLETELY DEMILITARIZING OUTER SPACE

Purging outer space of all military targets would be the most comprehensive arms control regime for outer space. But a core difficulty is that satellites and their activities cannot be neatly categorized as military or nonmilitary. Whether a navigation, weather, or communications satellite uses space peacefully could depend from moment to moment upon who was using its data and for what purposes. Moreover, it is hard to make the case that all military uses of space are undesirable. Military satellites may actually promote stability in crisis situations by reducing uncertainty, thereby reducing pressures for precipitate action and escalation. They facilitate close political control over distant military forces and can reduce the opportunities for third parties to do mischief. Since the data and satellites used for these desirable activities could also be turned to aggressive military purposes, it is impossible to ban the undesirable without curtailing the desirable as well.

Any arms control measures for outer space will encounter knotty verification problems, but a thorough ban on nonpeaceful uses may have more problems than can be tolerated. The sides would need to verify that all satellites launched neither performed forbidden functions nor carried banned equipment in an inert state, to be activated in a crisis. They would need to verify that data from civilian satellites would not be diverted to military purposes. They would need to verify that prohibited satellites had not been covertly stored and readied for launch in a conflict. This is a formidable, perhaps impossible, array of tasks, even if the sides were to agree upon intrusive prelaunch inspections of space payloads.

The impact upon American security of limiting space to nonmilitary uses would be substantial. Deprived of the use of space even in peacetime for routine military communications, navigation, and meteorology, the United States would have to turn to expensive and less effective ground-based alternatives and probably to a radical restructuring of military forces and commitments. Overseas bases for ground-based systems would have to be secured, with all that entails in new foreign-policy obligations and vulnerabilities. Unavoidably there would be opportunity costs for such alternatives, and if military spending in other categories were diverted to pay for gound-based alternatives, the result could be both fewer and less effective military forces.

If limiting space to nonmilitary uses impaired Soviet military forces to an equal or greater degree, however, U.S. security might still be served. Here, asymmetries in routine dependence upon satellite systems between the United States and the Soviet Union become crucial. Since the bulk of Soviet military forces are located in or near Soviet borders, their reliance upon satellites for military purposes is lower, and alternative Soviet ground-based systems have been maintained. This pattern may be changing as the Soviets build up an open-ocean navy and extend their military presence beyond Eurasia. But the core asymmetry will surely persist: The regions of vital concern to the United States (Europe, the Persian Gulf, Japan) are around the periphery of the Soviet Union, where Soviet forces can operate from home bases while the United States must bring power to bear at the end of long lines of logistics and communication. Thorough demilitarization of outer space would only emphasize this asymmetry and make the balance of military forces even more precarious for the United States.

PARTIALLY DEMILITARIZING OUTER SPACE

One way to escape some of these problems might be to regulate military satellite activities on a case-by-case basis, rather than ban them all. If on balance it seemed wisest to tolerate photo satellites but to ban radar surveillance satellites, then a bargain would be struck by the sides. The resulting provisions might defy categorization and might seem riddled with inconsistencies. But this approach would allow the sides to constrain the satellite activities that concern them most, and thus could go just as far toward removing the incentives for ASAT competition as a treaty with more elegant and comprehensive provisions. Additional provisions could be incorporated later to regulate new technologies or to accomodate new signatories.

It is conceivable that the two sides might agree to give up different kinds of satellites. More likely, prohibitions would apply symmetrically to both sides. Under this approach, therefore, the United States would have to relinquish those satellites that bother the Soviets and are the prime targets of the Soviet ASAT, and it would relinquish future rights to deploy American counterparts of the Soviet satellites that are targets of the U.S. ASAT (such as an American radar reconnaissance satellite).

It is difficult to judge the impact upon U.S. security of such partial demilitarization of space without knowing which satellites the sides plan to shoot at and therefore might insist upon outlawing. The history of the Soviet ASAT program does not help us infer much about the Soviet target list. Soviet ASAT tests reportedly began in 1968, with some precursor tests as early as 1966—implying a program decision made in the early 1960s. This suggests that the first U.S. photo surveillance satellites in 1960 may have provoked the Soviet program. But the Soviet ASAT is an omnibus weapon that has reached altitudes of 2,400 kilometers in tests and carried out intercept attempts against targets as high as 1,700 kilometers. Consequently, all U.S. low-orbit satellites could be potential targets, including all photo surveillance satellites and some ELINT, meteorology, and early navigation satellites. All this would suggest that the Soviet target list could be quite long, implying that a great variety of U.S. satellites would have to be banned before the Soviets would give up their ASAT.

On the other hand, Soviet ASAT tests have been relatively rare, compared to test programs for other weapons (only twenty ASAT tests over fifteen years, contrasted with dozens of test and training launches annually for Soviet ICBMs and SLBMs). All the tests have been at orbital inclinations (62–66 degrees) little used by U.S. miliatry satellites, and all interceptors have been launched from only one test range. These test practices may be due more to safety and political considerations (such as not having the booster fall on populated or foreign territory, or not launching the interceptor directly over the North Pole at the United States) than to performance limitations of the interceptor. On the other hand, they may also imply a short target list and a certain lack of urgency to the ASAT mission in Soviet military planning.

Supporting evidence is provided by the two treaties on space weapons submitted to the United Nations by Moscow. The first, proposed in August 1981, did not actually require the Soviets to dismantle their ASAT.[2] But further tests of their interceptor apparently would have been banned, and the use of ASATs against the satellites of other nations would have been prohibited, so long as those satellites did not carry "weapons of any kind". Since no current U.S. satellites carry weapons, as the term is conventionally understood, it appears that the Soviets were willing to impose limits upon ASAT testing without requiring any changes or sacrifices in *current* U.S. military satellite activities.[3]

In a second draft treaty, proposed in August 1983, the Soviets offered to dismantle their own ASAT in exchange for a ban on "any space-based weapons intended to hit targets on the Earth, in the atmosphere, or in space." Again, this suggested that the United States would be able to keep its own satellite programs and even duplicate Soviet satellite activities under Soviet ASAT arms control formulations. The Soviets did not offer to give up any of *their* current satellite activities as part of their proposal.

And what might the United States insist upon banning under "demilitarization"? The targets most often cited for the U.S. ASAT are the Soviet radar and electronic ocean surveillance satellites (RORSATs and EORSATs) whose purpose is to aid Soviet attacks on U.S. naval forces. Some characteristics of the American ASAT are certainly consistent with this mission. Currently the RORSATs and EORSATs orbit at an inclination of 65 degrees and at altitudes of about 500 kilometers. The U.S. ASAT, deployed on F-15s and operating out of bases in Tacoma, Washington, and Hampton, Virginia, could be readily placed under the ground tracks of Soviet ocean surveillance satellites, catching even newly launched replacements on their first or second orbits.

On the other hand, the ASAT system will be composed of two squadrons of F-15s (28 aircraft), presumably supplied with at least one ASAT per aircraft.[4] This number far exceeds Soviet ocean surveillance satellites – even allowing for ASAT failures, newly launched Soviet replacement satellites, and expansion of the RORSAT/EORSAT network during crisis. According to the most recent formulation of U.S. policy, the "primary purposes of a United States ASAT capability are to deter threats to space systems of the United States and its Allies and...to deny any adversary the use of space-based systems that provide support to hostile military forces." It has been publicly stated that the Joint Chiefs of Staff first drafted an ASAT target list in May 1978 and that the list was long enough to be divided into priority levels – highest priority targets were those Soviet satellites that "pose a direct, time urgent threat to U.S. forces."[5] Reportedly the 1978 list included forty-four low-altitude Soviet satellites, but the number has since been expanded and will grow even more by the time the ASAT becomes operational. When fully operational, the ASAT supposedly must be able to destroy no less than 25 percent of some 68 high-priority low-altitude Soviet satellites.[6]

If the United States demanded that all these Soviet satellites be removed under an agreement on partial demilitarization, it is hard to believe the Soviets would accept. The United States certainly could not accept a comparable ban on all such satellites of its own. This leaves open questions of whether such a long list is, in fact, indispensible for U.S. security and whether U.S. ASAT attacks are the most effective way of coping with threatening Soviet satellites. Those issues will be taken up in a moment.

Proposals for partially demilitarizing outer space could also stumble over many of the same obstacles as an attempt at thorough demilitarization. Even a shortened list of prohibited satellite activities must still be adequately verifiable, thus resurrecting the matter of intrusive inspection procedures. At the moment, satellites regarded as most threatening probably do have distinctive characteristics observable from a distance (such as large optics or antennas, indentifiable radio-frequency signals, distinctive orbital parameters, and so forth). There is no guarantee, however, that either the definition or the characteristics of the most threatening satellites will remain unchanged and susceptible to verification from a distance.

New satellite functions could also pose problems. If the sides were required to agree upon regulations governing them before new satellites were deployed, then the accord would be turned on its head. In effect, it would become an agreement on permitted (not banned) satellite activities, and, as such, it would grant each side a veto over all future space programs of the other side. On the other hand, if new satellite functions were permitted until regulations covering them could be negotiated, then incentives would remain for each side to acquire ASATs as a hedge against failure to reach agreement on new technologies.

Perhaps more boldness and ingenuity would cut through these complexities. An ingenious if radical scheme for coping with satellite activities that can be either "desirable" or "threatening," depending upon the user's intentions, might be to place such satellites under international or bilateral control. The French proposal in 1978, to establish an International Satellite Monitoring Agency as an independent arms control verification and crisis monitoring service, offers a suggestive framework for international control of surveillance satellites. Proposals for a jointly-manned U.S.–Soviet crisis center might provide the foundation for bilateral supervision of "desirable" satellite activities. And a bold scheme for verifying that satellites were demilitarized might provide for prelaunch inspection of space vehicles.

Boldness and ingenuity do not invariably make for effective and negotiable arms control, unfortunately. Schemes for international or bilateral supervision of satellites must tackle the matter of procedural safeguards on the collection and dissemination of satellite data. If the data genuinely is the sort that could be turned to threatening purposes, then the only safe procedure would be that all such data must be denied to the United States and Soviet Union entirely or granted to them equally. In the first case, collection and evaluation would have to be done entirely by neutral third parties. From the U.S. perspective, this would mean that decisions on matters touching the vital security of the United States (such as, are the Soviets cheating, are allies about to be attacked, and so forth) would be made by an agency that is neither politically nor legally responsible for U.S. security, nor likely to bear the brunt of its own mistakes. A bilateral arrangement that shared all regulated satellites and data equally, on the other hand, might simply benefit the Soviet military, since the quality of U.S. satellite data is probably higher than that of the Soviets. Neither arrangement could ensure that prohibited satellites would not be covertly acquired and launched anew during confrontations.

Verification schemes that would require prelaunch inspection of space vehicles will not be acceptable unless the sides have already agreed to thorough demilitarization of space. If only some military satellite functions are banned, the sides will insist upon some secrecy at launch sites, to protect the effectiveness of functions that are not banned. Any inspection system that exempted some satellites or some launch sites from scrutiny would allow a certain amount of cheating; certain areas (or launches) in effect would become sanctuaries for evasion. Inspections by neutral third parties are unlikely to be satisfactory to the United States or the Soviet Union. Neither neutrality nor responsibility could be assured, and the technical expertise indispensible to effective inspections would be difficult to find in genuinely neutral countries that lack sophisticated space programs of their own.

The prospects of demilitarizing outer space, fully or partially, are not very impressive. But the same security concern that argues *against* strict demilitarization—that is, the U.S. need for military satellites—also argues *in favor* of reducing the Soviet ASAT threat to those satellites. This does not necessarily require an arms control agreement. The United States could conceivably reduce the Soviet ASAT danger by a combination of satellite survival measures that made U.S. space systems less vulnerable and by the threat of ASAT retaliation. Essentially, this is the policy adopted by the United States thus far. This policy should be weighed against the arms control alternative. Arms control limits on antisatellite weapons could take several forms, ranging from a ban on all use, testing, and deployment through a ban on use alone.

BANNING ASAT USE, TESTING, AND DEPLOYMENT

The most comprehensive arms control measure for ASATs would prohibit all testing, deployment, and use of any weapon to damage or destroy satellites. This approach would resemble, in general terms, what the United States proposed to the Soviet Union during bilateral ASAT arms control negotiations in 1978 and 1979. In broad terms, it would also resemble key provisions of the Soviet draft treaties submitted to the UN General Assembly in 1981 and 1983. Deployment and testing of weapons permitted by other agreements (such as ground-based ABM interceptor missiles) could not be banned, even though such weapons have residual ASAT capabilities. Testing them against satellites or in any other ASAT mode would be prohibited, however. Under this approach, the United States would gain protection for its own satellites, though at the price of forgoing its own ASAT.

And what would the United States be giving up? The missions assigned to the U.S. ASAT, as noted previously, are reportedly to deter Soviet ASAT use through threat of a tit-for-tat response; to attack Soviet satellites that threaten American military forces; and perhaps to intercept the Soviet ASAT in flight as a means of protecting U.S. satellites.

Although it is sensible for the Defense Department to consider, as alternative options, both attacking Soviet satellites with an American ASAT and deterring the Soviets by withholding such attacks, it would be difficult to perform both missions simultaneously. The United States would need truly awesome ASAT capabilities and satellite defenses if it hoped to deter Soviet ASAT use, even while the U.S. Air Force destroyed selected Soviet satellites. Every Soviet satellite destroyed would also be one fewer held "hostage" in support of ASAT deterrence. Deterrence would require that such hostage satellites be of great value to the Soviets but pose only tolerable threats to U.S. military forces, and be of greater benefit to the Soviets than what they would gain by attacking U.S. satellites.

This last assumption is the one most often assailed by ASAT critics, who note that the Soviets could emerge better off from an ASAT exchange because they have a greater launch capacity for replacing satellites, more ground-based substitutes to perform key satellite functions, and greater likelihood of bringing those substitutes to bear because clashes between the United States and Soviet Union would probably occur nearer Soviet borders. If the progressive loss of satellites on both sides would work to the relative advantage of the Soviets, then the United States has little prospect of deterring Soviet ASAT use by threats of retaliation, either before or during conflict.

Neither is it clear that Soviet satellites held hostage would be of great value to the Soviets, while posing only tolerable threats to the United States. In wartime, the Soviets are likely to prize satellites precisely to the degree that they bolster Soviet ability to threaten U.S. military forces. Moreover, the United States cannot know with surety which satellites the Soviet leadership would most fear losing; Soviet leaders themselves may not know until they see a confrontation unfolding. Consequently, to avoid inadvertently "killing the hostage," the United States would have to refrain from all attacks on Soviet satellites. In order to make ASAT deterrence work, then, the United States must forgo use of its own ASAT weapons, leaving U.S. forces to protect themselves by other means against threatening Soviet satellites. If deterrence breaks down and the United States decides to destroy Soviet satellites, U.S. forces will still need protective measures over the hours or days that it takes the U.S. ASAT to destroy all threatening Soviet satellites (including newly launched replacements). If the Soviets start a conflict, an ASAT system may afford the United States no protection at all, since the Soviets will have been able to exploit their satellite capabilities to target and coordinate their military forces up to the moment battle begins.

Would not an ASAT provide a cushion of confidence and a useful option for the United States in some cases? The difficulty is that in order to have genuine choice among courses of action, the United States must purchase the requisites for each option. If the United States opts for ASAT deterrence when conflict begins, then all the defenses that its military forces require to survive Soviet satellite threats must already have been bought and deployed. Conversely, if the United States decides to initiate ASAT attacks, then its own satellite functions must already have been made survivable against Soviet ASAT retaliation. An American ASAT system will compete for funds with both of these defense requirements but cannot itself make up for weaknesses in either. An ASAT held back for deterrence cannot help an aircraft carrier in the Persian Gulf blind a Soviet RORSAT or shoot down Soviet antiship missiles; an ASAT launched against Soviet RORSATs cannot protect U.S. photo surveillance satellites from Soviet ASAT attack.

The tradeoffs here are not trivial. The cost of the U.S. ASAT system is currently pegged at $3.6 billion, equal in cost to a Nimitz-class nuclear-powered carrier or 200+ F-16 fighter aircraft. The General Accounting Office, however, projects that the ASAT system's final cost might well run to "tens of billions."[7] At that level, it would rival what the Navy and Air Force each plan to spend over the next five years on *all* new fighter aircraft. Shifting the funds and doubling the number of Navy or Air Force fighter squadrons purchased might well protect U.S. forces better than buying an ASAT.

An American ASAT could contribute directly to U.S. satellite survivability if it were able to intercept the Soviet ASAT in flight. But Soviet options for thwarting such intercepts are several. They could devise decoys to accompany their ASAT as it passed over the U.S. ASAT bases, or they could exploit their ASAT's capacity to maneuver. The U.S. space-tracking network can track the Soviet ASAT quite accurately, but it cannot keep it constantly in sight because the number of tracking sites and their fields of view are limited.[8] For example, if the Soviets were to launch their ASAT northward over the Pole, in pursuit of a U.S. satellite in an orbit 300 kilometers high, the ASAT would come within line-of-sight of U.S. radars in Alaska or Greenland about ten to fifteen minutes after launch, while the ASAT was still about 1,800 kilometers away. But if those radars have an angle of view, from horizon to zenith, of only 20–40 degrees, then the ASAT will pass out of their view within a minute or so, while the ASAT is still a 1,000 kilometers or so out in front of the radars. Depending on its path, the Soviet ASAT may come within line-of-sight of a U.S. tracking radar in North Dakota about five to seven minutes later, again passing out of view a minute or so afterwards. Otherwise, about twelve to fourteen minutes will then go by before the Soviet interceptor would pass over the U.S. ASAT base in Washington or Virginia. The Soviets could program their ASAT to maneuver just after it passed out of radar view, so that any information passed from the radar to the U.S. F-15/ASAT would be instantly obsolete. The United States would be forced to reposition its F-15/ASATs so they could attack while the Soviet interceptor was in radar view. This would truncate the time available for the United States to pinpoint the Soviet ASAT's orbit and calculate an intercept point—perhaps defeating the attempt.

The Soviets could also loft their ASAT into an elliptical orbit that might carry it above that reach of the U.S. ASAT while passing over the United States; it would then close in on its target somewhere in sight of the Soviet Union. These are not foolproof tactics, and the very mobility of the American F-15/ASAT gives it some flexibility in finding the most favorable point for intercept. Yet to exploit that mobility fully, the United States would have to invest in overseas basing, carrier-based ASAT aircraft, and a more extensive space-track radar network—ventures that could add billions of dollars to ASAT costs. And if the Soviets develop their own direct-ascent ASAT, which could hit American satellites directly over Soviet territory, then the investment would have been wasted.

Even if the U.S. ASAT does not prove to be an especially good "defensive" weapon for deterring or intercepting Soviet attacks on U.S. satellites, it would still have a third mission—destroying Soviet military satellites. In weighing the ASAT's effectiveness for this mission, it must be remembered that the tactics just mentioned that might allow the Soviet ASAT to elude attack could, in principle, also be exploited by other Soviet satellites. And if their satellites cannot be readily equipped to defend themselves, the Soviets could hope to saturate and exhaust the U.S. ASAT system by constantly lofting new satellites to replace those destroyed. If the United States were able to attack these replacements on their first orbital pass, the Soviets would waste a lot of equipment without even getting the benefit of one orbit's worth of data. But the game could eventually be theirs.

ASAT attacks are not the only technique for thwarting Soviet satellites. Other options are available to the United States and might be less inflamatory during crisis and more difficult for the Soviets to cope with: radio silence and jamming for ELINT/SIGINT satellites; jamming and radar decoys for RORSATs; camouflage, decoys, and night maneuvers for photo satellites; jamming of satellites or ground receivers for communication and navigation satellites. Again, none of these is foolproof, and some techniques that rely more upon finesse than brute force may require reasonably good guesses about the technical design of Soviet satellites. On the other hand, for reasons mentioned above, U.S. military forces must be equipped with these techniques even if the United States has an ASAT, either to support ASAT deterrence or to protect themselves until all threatening Soviet satellites are destroyed. So what is at issue is not whether such techniques should be explored and deployed but how much effort should be put into them rather than into an ASAT. These other techniques have the advantages that they can be shared with allies and many could be used in cases where the Soviets might share satellite data with their own clients while formally staying out of a conflict themselves. Moreover, these techniques are reversible countermeasures that could be selectively halted if it proved desirable to restore Soviet satellite functions, such as communication with troops, at some point in a conflict.

In sum, there are reasonable grounds for concluding that the United States might not be giving up much if it agreed to forgo ASATs. However, this does not clinch the case for an ASAT ban. If the primary U.S. security interest in space arms control is to protect American satellites, the United States would not gain much unless assured that the Soviets, too, had given up their ASAT capability. Verification again is the key.

In any arms control accord, there are two verification standards that could apply: adequate and absolute. An accord might be judged adequately verifiable if cheating could be detected before it became militarily significant, even though some lesser cheating might go undetected. On the other hand, a treaty is supposed to be a solemn obligation, and therefore any and all violations ought to be prevented through strict verification. This absolute standard, if applied rigidly, would put an end to virtually all international agreements, between friends or foes, since not even the most intrusive of inspection measures could guarantee that not one single act of cheating had taken place. Yet much harm can be done by ignoring this stricter standard, since a pattern of niggling violations does constitute a political insult, erodes public confidence in arms control, and may encourage more serious attempts at evasion.

Even by the standard of adequate verification, a total ban on ASATs would face harsh judgment from some skeptics. It has been argued that vital American military satellites are so few in number, and their loss so potentially disabling, that the United States could not tolerate any gaps in verification that would allow even residual or improbable ASAT threats. Hence, even verifying that all existing Soviet ASAT interceptors had been dismantled would not be adequate, because the United States could not be assured that they would not be covertly manufactured again. Moreover, the Soviets would retain residual ASAT capabilities in other weapons and space vehicles. Nuclear-armed ICBMs could be used against satellites at all altitudes, and ABM interceptor missiles could reach satellites in lower orbits. Laser weapons, whether ground- or air-based—even if only research and development models—could potentially be used to blind photo sensors on satellites at high altitudes or physically damage low-orbit satellites. Manned Soyuz vehicles, Progress resupply vehicles, even a future Soviet space shuttle could rendezvous with and damage American satellites in lower orbits. Since many of these things could not be banned without crippling peaceful uses of space, it seems that a strict ban on ASATs faces insurmountable verification obstacles.[9]

While there is some validity to these arguments, it is also true that the vulnerability of current U.S. satellites to even crude ASAT technologies is a consequence of thoughtless policies in the past that the United States can no longer follow, even in the absence of ASAT arms control. In the past, tight budgets, high launching costs, and limitations on booster payloads encouraged satellite designers to cram multiple functions and as much equipment as possible onto relatively few satellites, with little attention to survivability measures—even though the Soviet ASAT threat has existed since 1968. Presidential directives, in both the Carter and Reagan administrations, reversed this trend, and satellites now being deployed have various defensive measures incorporated in them. Given the long lifetimes of U.S. satellites, and hence slow replacement rates, it will be five to ten years before all vital satellites are protected by such features. Fortunately, the satellites with the lowest replacement rates are at higher altitudes and thus currently are beyond reach of both the Soviet ASAT interceptors and most residual ASAT threats.

A ban on possession of ASATs, even if not strictly verifiable, would be strengthened by an accompanying ban on ASAT testing. The United States knows the characteristic test pattern of the current Soviet ASAT quite well, and the Soviet Union is ringed with U.S. surveillance systems. The United States presumably also knows Soviet non-ASAT space programs quite well, so that attempts to disguise covert ASAT tests as some other space activity would risk detection. Since the Soviets must monitor their own tests and recover data from test vehicles, they may be forced to conduct covert tests directly over Soviet territory, where U.S. verification systems are focused. Military skeptics of arms control occasionally argue that *the other side* can build highly reliable weapons covertly without full-scale testing; yet few military planners would risk *their own* operations on weapons that had never been tested in realistic simulations. Over time, in the absence of tests, the Soviets might come to doubt the reliability of a covertly acquired ASAT or its capacity to cope with newly deployed U.S. satellite survivability measures.

The pertinent question, then, given that the United States must take measures to make its satellite functions less vulnerable anyway, is whether protecting satellites would be easier with arms control limits on ASATs or without. Any limitations, even ones where verification confidence was less than ideal, could potentially be useful if they narrowed the variety of ASAT threats and slowed the pace at which new ones were developed. In some measure, even limits that channeled competition and conflict in one direction rather than another might have value for U.S. security. For instance, no ASAT agreement could completely remove the residual threat to satellites posed by Soviet nuclear-armed ICBMs and ABM missiles. But firing off dozens of nuclear-armed missiles at American satellites surely would be a grave step that the Soviets would be reluctant to take in any confrontation short of nuclear war. Lofting a conventional ASAT at U.S. satellites, on the other hand, might be regarded by the Soviets as less grave, though the consequences for U.S. military effectiveness might be the same. Hence, a ban that eliminated conventional ASATs might be useful, even though the nuclear ASAT threat remained.

Due regard for the second, "absolute" standard of verification would argue against becoming too casual here, however. Under a total ban on ASATs, the United States, at a minimum, would have to demand fairly rigorous verification measures to ensure that the current Soviet ASAT system was dismantled, and this could prove nonnegotiable with the Soviets. A core problem is that the Soviet ASAT shares a common booster rocket (adapted from the old SS-9 ICBM) with the RORSAT and EORSAT programs, and since all are launched out of the Tyuratam missile range, they probably also share common launch and control facilities. In addition, a related booster has been used to launch several dozen satellites out of the Plesetsk missile range since 1977. The Soviets may reject a demand that they dismantle all these facilities. And the United States might not wish to have the Soviets switch all these programs to other boosters and launch facilities, since that would arouse suspicions that the ASAT could be covertly reacquired and switched in the same fashion. So the matter of how much to demand from the Soviets under dismantling poses a dilemma for the United States: Demanding too much risks rejection; too little may offend public sensibilities about verification.

If the risks, uncertainties, and dilemmas of a thorough ban on

ASAT weapons seem too overwhelming, there are other, less sweeping arms control formulation.

BANNING ASAT SPACE TESTING

An ASAT test moratorium could curtail ASAT competition, reduce the pace (and perhaps the extent) of ASAT modernization, and inhibit the use of ASATs by raising doubts in the minds of military planners about the operational reliability of untested weapons. (Testing in this context would mean tests in space or against objects in space.) A ban on ASAT tests alone would not offer U.S. satellites as much protection as a thorough ban on tests and possession, but it might be more verifiable and, because the United States would retain its own ASAT, remaining gaps in verification might be regarded as more tolerable.

The intended effect of a test ban, in reducing the ASAT threat by undercutting reliability, is perhaps demonstrated by Soviet experience in the early 1970s. The Soviets did halt their ASAT tests for over four years between December 1971 and February 1976. Prior to that hiatus, all tests had involved intercepts of a target on the ASAT's second orbit, and of the seven tests conducted, apparently about 70 percent were successful. Between February 1976 and June 1982, thirteen more tests were run. This new tests series involved intercepts on both the ASAT's first and second orbital passes, and a new homing sensor was reportedly also tested. The success rate apparently dropped to less than 50 percent, including a string of failures with the new sensor.[10] If even modest changes from prior test practices and hardware produce such deterioration in performance, it would give the Soviets something to ponder when they contemplate going after U.S. satellites in orbits never attempted by their ASAT in tests, or if they were to attempt to upgrade their ASAT covertly, without full testing.

The Soviets in fact offered an ASAT test moratorium in August 1983, simultaneous with—but evidently separate from—a proposal that all ASATs be dismantled.[11] The timing was opportune for the Soviets, since test launches of the American ASAT from an F-15 had not yet begun. In the United States, there were those who argued that once U.S. tests were conducted, the Soviets would lose interest in ASAT arms control and would insist instead upon pushing for a more modern ASAT of their own.

There are major differences between the U.S. and Soviet ASAT weapons that would be frozen by a test moratorium, if it actually impeded further modernization. The U.S. ASAT is a product of 1980s American technology; the Soviet ASAT interceptor of 1960s Soviet technology. The U.S. system is air launched and can be carried by F-15s to advantageous places for launch against prospective targets; the Soviet system must use fixed launch pads and thus must wait for the orbits of targets to pass overhead. The U.S. ASAT has a small booster (1,000+ kg), and the F-15 can be readily reloaded; the Soviet ASAT booster is massive (90,000+ kg), surely more expensive, more cumbersome, and more demanding of its launch facilities and crews, with undoubted limits on its reload rate. In general, the U.S. ASAT system would appear to be more flexible and expandable, so the Soviets might object that a test moratorium—once the United States has tested its ASAT against space targets—locks them into an inferior position.

Such complaints might not be serious. The pertinent issue is not whether the Soviet ASAT is "as good as" the American, but whether theirs is "good enough" for their missions. That the Soviets have persisted with their current interceptor may indicate that they are content with its capabilities for the missions of interest to them. On the other hand, reported Soviet tests of a new homing sensor on their ASAT perhaps hint at a gap between the interceptor's performance and Soviet operational requirements. And while the Soviet ASAT may be good enough against current targets, it may lose that edge as the United States improves its satellites' defenses.

An unknown factor is how the Soviets view the future. Clearly they have security requirements that go beyond competition with the United States in ASAT weapons, and thus they must weigh opportunity costs, just as the United States must. In addition, if the American ASAT interceptor proves as nimble as its designers intend, the United States has a chance of moving ahead of the Soviets in potential high-altitude ASAT capability in the near term, albeit temporarily. (Since the U.S. ASAT is so small, it could more readily be matched with relatively inexpensive, available boosters.) The Soviets might believe that the relative balance of advantages in an ASAT duel is quite scenario-dependent, and they are in a better position to judge which scenarios bother them most and how rapidly they will be able to close a temporary U.S. technical advantage.

A more substantial problem for the Soviets is whether a test ban would constrain the U.S. ASAT. Given the small size of the American interceptor and its booster, the rapidity with which it could attack targets, and its mobility, the Soviets could be hard pressed to verify a test ban using their space surveillance system. But then, American society is much more open, and deliberate violations of an ASAT agreement would be hard to conceal. However, the American ASAT is derived from technology originally intended for the Army's anti-ballistic missile defense program. Related technologies are still part of the Army's "overlay" ABM project, and the Soviets might suspect that ongoing tests of the ABM system were, in effect, proof tests of the ASAT interceptor. The Soviet ASAT, on the other hand, does not appear to be similar at all to any other weapon or space program and thus is a poor candidate for such circumventions of a test ban.

Clearly, thorough-going bans on possession or testing of ASATs pose problems that, if not insurmountable, are at least daunting. The question arises, then, whether there might be other less comprehensive but still useful arms control formulations.

BANNING "NEW TYPES" OF ASATS

This formulation would concede the difficulty of limiting conventional ASATs, given that the Soviets have an operational system, the United States has begun tests of its ASAT, and both sides' systems, if banned, could perhaps be reconstituted on short warning in the future in abrogation of a total ASAT ban. A "no new types" prohibition would ban all ASATs except the existing American and Soviet interceptors, and significant alteration or modernization of these existing systems would also be banned.

The purpose of a "no new types" limitation would be to confine the conventional ASAT threat to lower-orbit satellites only. The maximum altitude reached by the Soviet interceptor in tests thus far is about 2,400 kilometers; its full altitude capability may be two or three times that, against some targets. The altitude reach of the U.S. ASAT, launched from an F-15, has not been publicly announced. However, the Soviet satellites that reportedly are its prime targets have orbits of at least 500 kilometers, so the altitude capability of the U.S. ASAT is at least that, and presumably quite a bit more. The capability of either interceptor is not a single figure, of course, but varies according to such factors as launch location, the direction of launch, required final maneuvers, and so on. Even with these variances, a "no new types" limitation would confine the conventional ASAT threat to the lowest several thousand kilometers of space.

Although the first order of business would be to limit the modernization of the ASAT interceptors themselves, limitations could be extended to the number of launch sites, storage capacity at launch sites, number of tests annually, and so forth, in order to further constrain the Soviet ASAT threat. The U.S. ASAT system is not a very good candidate for some of these restraints, however, since its "launch site" will be an F-15 and it will be small enough to be hidden at any airbase. So such limits could be considered, but they pose obvious verification problems for the Soviets.

A "no new types" approach to ASAT arms control concedes a great deal to the perceived security concerns that have motivated the American ASAT program. The United States would retain the capacity to threaten Soviet satellites in lower orbit, including the RORSATs and EORSATs. It would simultaneously grant something to strategic stability and arms control. Many American and Soviet military functions that can reduce tensions during confrontations—such as attack early warning, communications with strategic retaliatory forces, detection of nuclear detonations—are carried out by satellites at higher altitudes, of 20,000 km or more. A "no new types" limit on ASATs would enhance stability by offering protection to these satellites. In general, defending all satellites, high altitude and low, could become more manageable for the United States, since a "no new types" agreement would freeze the Soviet ASAT offense while leaving U.S. satellite defense technology free to advance. This approach could also shrink or even close a potential ASAT loophole in the ABM Treaty, by banning sophisticated ASATs (such as space-based lasers) that could be prototypes of orbiting ABM systems.

ASAT arms control schemes generally pose tough verification problems, and this one is no exception. The key issues are two. First, could permitted ASATs be readily modified in some way to grant them significantly higher altitude capabilities? Second, could other, prohibited ASATs with greater capabilities be acquired covertly?

Some additional capability could conceivably be squeezed covertly into the existing Soviet ASAT by increasing the fuel capacity of the booster and/or interceptor. But this would yield little advantage. The velocity increments necessary to loft the interceptor higher diminish as altitude increases, thus seeming to reward marginal cheating. But the fuel to grant that added velocity must itself be transported up to altitude, and this cascades back into an almost exponential increase in fuel requirements and into potentially verifiable increases in booster size, and so forth. The Soviet ASAT has been tested up to 2,400 kilometers, high enough to reach all low-orbit U.S. satellites. But the next set of important American satellites, the GPS/IONDS navigation and nuclear-detection network, is ten times higher (at 20,000 km), quite beyond the reach of such marginal improvements in the Soviet ASAT.

For the same reasons, the Soviets would gain little by cheating and configuring other non-ASAT systems for ASAT attacks. Most of these systems have inherently limited ASAT capabilities and thus would not add to the threat, even if not rigorously verifiable. As noted above, ground- or air-based lasers could blind unprotected optical sensors at higher orbits, but otherwise they could do physical damage only to the lowest satellites. The altitude reach of ground-based ABM interceptor missiles is also limited to several hundreds of kilometers. Space mines maneuvering alongside low-orbit satellites would immediately raise suspicions and would be vulnerable themselves to low-orbit ASATs. The Soviet Soyuz and Progress space vehicles are even more massive than the Soviet ASAT and require a much larger booster, with longer preparation time, to reach even low orbits. The very fact that a "no new types" accord would permit the Soviets to keep their existing low-altitude ASAT would diminish both the incentive and the advantage in such cheating.

A greater concern is that the existing Soviet interceptor would be covertly mated with a new, more powerful booster capable of reaching high altitudes. Space vehicles are highly complex things, however. How readily or rapidly the intricate interface between an interceptor and a new booster could be perfected without testing is a matter of dispute among specialists. There is a nice asymmetry between the U.S. and Soviet cases here. It would be hard for the United States to feel confident that the Soviets were not covertly engineering the necessary interface. But the current Soviet interceptor is so massive that it would require an enormous booster, comparable to the Soyuz or Salyut rockets and requiring hours or days for launch preparations, to attack higher satellites. The smaller U.S. ASAT, on the other hand, would require only a modest, readily available booster (such as a Minuteman ICBM) to reach geosynchronous orbits, yet the openness of U.S. society would make such covert preparations very difficult. Even if the Soviets could loft their interceptor up to higher orbits, it might not function properly. High-orbit attacks place greater demands upon guidance, homing, temperature regulation, battery power, and maneuvering systems than low-orbit intercepts, and it is not clear that the Soviet ASAT was designed to meet these higher requirements.

An illicit high-altitude ASAT could be pursued by the Soviets in other ways. They might try to develop an entirely new ASAT—an orbiting laser perhaps, or a miniature ASAT interceptor similar to the American ASAT. Nuclear-armed ICBMs, launched to the general vicinity of a passing satellite and detonated, are another approach. These would breach the nuclear threshold and could damage Soviet satellites as well, but they assuredly could kill American satellites up to the highest orbits. One problem encountered in launching ASATs from earth against high-orbit satellites, however, is that travel time to geosynchronous orbits may be six hours or more, giving ample warning of attack. Space mines are, therefore, a more inviting form of high-altitude ASAT, since they could be placed near their targets well in advance and exploded upon command. But to avoid suspicions, a Soviet space mine would have to conform to current international practices and remain hundreds of kilometers away from its target at geosynchronous orbits. It would then require sophisticated homing and maneuvering capabilities to attack upon command. A nuclear warhead on a space mine could kill its target from such ranges, if the warhead had a near-megaton yield. But such nuclear warheads deteriorate and become unreliable unless replenished and maintained regularly.

Doing many of these things without space tests, or concealing the tests from U.S. verification systems, would be both technically and politically risky. Again, the tests must be conducted in places and ways that allow the Soviets to recover test data. Moreover, both ASATs and targets must initially be launched from earth. Thus, many aspects of the tests would be exposed to the phalanx of U.S. monitoring systems deployed around and over the Soviet Union.[12]

Nevertheless, the current U.S. difficulty in monitoring space activities and objects at very high altitudes is a complication for a "no new types" ASAT accord. NORAD, the North American Aerospace Defense Command, maintains and constantly updates its Resident Space Object Catalogue to record the identity and orbits of all space objects at all altitudes. NORAD's space tracking radar network has recently been upgraded to improve tracking accuracy and to close gaps in coverage. Upgraded radars in the United States, Turkey, and the Pacific will extend radar tracking out to geosynchronous orbits. Still, the reach of ground-based radars, their field of view, and their ability to resolve details is limited. When completed, the five tracking sites of the American GEODSS (Ground-based Electro-Optical Deep Space Surveillance) system will augment ground radars and will provide "precise, highly detailed optical and orbital signature data on man-made objects approximately 5,500 km from Earth and beyond."[13] But GEODSS will not reach full operational status until 1987, each site evidently will be able to survey no more than half of the sky that sweeps overhead each night, and treaty verification may require more than the type of data GEODSS can provide.

On the other hand, these verification problems should not be overemphasized. In the absence of ASAT arms control, the United States must still monitor the Soviet ASAT threat in order to design programs for satellite protection. American space surveillance capabilities will evolve and improve, including the ability to monitor events *in* outer space *from* outer space. During the first space shuttle flight, for instance, the United States demonstrated an impressive capacity to determine which thermal tiles on the shuttle were damaged, supposedly by focusing a photo surveillance satellite on the shuttle. The United States has also funded the SIRE (space infrared experiment) project to develop a satellite that will be able to detect and track other satellites.[14] So *future* ASAT cheating scenarios ought properly to be matched against *future* U.S. verification capabilities. The demands of arms control verification are admittedly somewhat more rigorous than threat monitoring (because of desires for "absolute" verification), but they are not so different. In preparing to do one, the United States will be preparing to do the other as well.

At first glance, this "no new types" approach to ASAT arms control appears to grant a little something to all the missions assigned to the U.S. ASAT. And that it does, for the near term. Over time, however, these ASAT limitations would presumably induce the Soviets to reshape their own satellite programs. In effect, a "no new types" accord would establish a sanctuary for Soviet satellites at altitudes beyond the reach of the U.S. ASAT. This invites the Soviets to redesign those satellites that are the probable targets of the U.S. ASAT, so that they could carry out their tasks from higher orbits. "Redesign" is not precisely the right term—entirely new satellites must be devised. The effectiveness of RORSATs, EORSATs, ELINT/SIGINT, and photo satellites declines by the square of the altitude or more, so that moving them up into orbits twice as high means enhancing their performance by a factor of four or more, just to stay even. In some instances, performance improvements of this magnitude may involve sensor technologies that the Soviets have yet to master. So while the most threatening of Soviet satellites will probably remain vulnerable for a good number of years, this may not be true forever. Eventually the United States could end up with an ASAT but few reachable targets (and the same could be true for the Soviets). In an evolutionary way, a "no new types" accord could closely approximate a "no ASATs" agreement. As such, it might ultimately yield all the security benefits and/or problems for the United States of a thorough ASAT ban.

BANNING ASAT USE

Under this formulation, the sides would agree not to use ASATs against each other's satellites. In addition, they could agree not to interfere with the normal functioning of each other's satellites by doing such things as jamming, blinding by lasers, issuing false commands that cause the satellite to malfunction, and so forth. And the sides could agree to establish "rules of the road" in space to minimize activities resembling ASAT attacks. This approach would not prohibit the possession or testing of ASAT weapons. Nor would it ensure the security of satellites in the event of conflict, since any prohibition on use or interference will be tenuous, if not accompanied by limits on the weapons themselves.

On the other hand, the merit of this formulation would lie in the reduced probability of provocative actions during peacetime and up through the outbreak of conflict. By specifying in a formal treaty the types of behavior in space that would be regarded as inflammatory (and presumably cause for reprisal), an agreement of this sort could reduce misunderstandings and in some degree deter hostile actions in space. In these respects, it would make a contribution similar to the U.S./Soviet Agreement on the Prevention of Incidents On and Over the High Seas of 1972, which has helped curtail petty but hazardous and provocative confrontations of American and Soviet ships and aircraft. One practice that that agreement specifically prohibits is simulated attack runs by one side's aircraft or ships against those of the other side—a dangerous practice that ought to be prohibited in outer space as well.

A useful feature of this approach is that it could be a multilateral undertaking, either in a newly negotiated treaty among interested states or as an extension of existing treaties. Limitations on certain forms of interference with the space activities of other states, for instance, are contained in both the Outer Space Treaty and the Conventions of the International Telecommunications Union. These obligations could be expanded in protocols to cover damaging, destroying, changing the orbit, or otherwise rendering inoperable the satellites of other states. "Rules of the road" and a ban on additional forms of interference could be appended in a similar fashion.

The precise form of "rules of the road" would be derived from the characteristics of existing or anticipated ASATs. For instance, the current Soviet ASAT starts its attack by going into an orbit within one degree or so of the target's orbital plane. So an appropriate rule of the road might be to prohibit any launch by one state into an orbit within, for example, two degrees of the orbital inclination and right ascension of another state's satellite. Similarly, the U.S. ASAT approaches its target on direct-ascent from any azimuth, so an appropriate rule might be that the path of one state's spacecraft during its first two orbits may not cross within, for example, one hundred kilometers of another state's spacecraft. To cope with possible space mines at higher orbits, minimum separation distances of several hundred kilometers (already maintained in order to minimize radio interference) might be enforced. Other rules of the road could be devised to cover Molniya orbits, direct ascents to geosynchronous, and so on. If a need arose to violate the rules for some peaceful purpose, consultation between the parties would be required.

Potentially, such restrictions on free passage through space could impose costs on non-ASAT space programs. They may narrow the "launch windows" for spacecraft, for instance, require extensive calculations of the orbits of other states' satellites when selecting windows, and force long postponements when a window is missed.[15] Such problems may be especially irksome because, in the final tally, neither rules of the road nor bans on use or interference can themselves prevent an ASAT attack. They can only make it more blatant. It should be borne in mind, however, that free passage has been and must be constrained in any case as space becomes more crowded, and thus far the users of space have been able to adjust satellite design and practices to live within imposed rules.

Verification of such "rules" could never be perfect, because the U.S. space tracking network cannot keep all satellites and all sectors of space in view constantly. So transgressions could occur, though it seems improbable that the Soviets could gain a worrisome military advantage by such violations. And whether there is an ASAT agreement or not, the United States will want to keep tabs on close approaches between all American and Soviet spacecraft. Because attacks on or interference with its satellites might be the precursor to other hostilities, the United States will have an interest in determining when its satellites are being tampered with and by whom. Impact sensors and other detectors for monitoring the nature and source of interference are being installed in any case. A ban on interference should make these detectors even more useful as warning systems since, without a ban, U.S. satellites would be vulnerable to repeated "spoofing" until a run of false alarms had reduced alertness to a genuine attack. Under a ban on such interference, any tampering would be a clarion demanding attention.

SUMMARY APPRAISAL

The visions of impending cataclysm that animate arms control efforts in the arena of nuclear weapons seem inappropriate in the realm of antisatellite arms control. Neither ASAT technologies nor superpower dependence upon satellites has yet reached the point where a space war in the near future would offer prospects of decisive military advantage. The dangers in space arms competition are future ones, and hence more speculative and less compelling. The dangers are that in some local confrontation between American and Soviet forces, one side and then the other will be tempted to better its position by striking at the other's satellites; that as the satellites networks through which the sides gain information and exercise control over events begin to collapse, escalation will follow.

An additional hazard of ASAT arms competition with the Soviet Union is that it promises to be very expensive and ultimately unrewarding for the United States, paid for by the diversion of energies, attention, and resources from other pressing military requirements.

It would obviously be easier for the United States to rein in its own

ASAT program if a verifiable agreement with the Soviet Union could be drafted. Doubts regarding verifiability are clearly a major obstacle. It is a nice irony that the U.S. ability to verify an ASAT agreement is now being rapidly improved by the Air Force's Space Command as it prepares itself for space warfare. The matter of verifiability ought properly to be assessed in light of the enhanced American monitoring capabilities that the Space Command is pursuing.

Finally, no arms control proposal can be an impenetrable shield, protecting the United States from all Soviet threats and from our own imprudence. This is both an unreasonable expectation and a peculiar standard for arms control. No matter what the formulation of an ASAT accord, some residual dangers to American satellites will persist, from other non-ASAT weapons and from nonmilitary space programs. Even under arms control, a healthy satellite defense program, substantial vigilance, and just plain good sense will still be indispensible.

What arms control could offer is some insurance against catastrophe when our good sense and vigilance fail us.

NOTES

1. See, for instance, the remark of Presidential Science Advisor George Keyworth: "[A space-based laser] may not necessarily be the best way for the ASAT mission, but a [laser] geosynchronous anti-satellite capability is important to test the [ABM] technology to destroy missiles." Quoted in *Aviation Week & Space Technology* (18 July 1983): 18–21. Along the same line, the Fletcher Panel established by President Reagan to make recommendations on "star wars" ABM research projects, reportedly proposed an early technology demonstration of ABM ground-based lasers, not in an antiballistic missile mode but as an antisatellite device. See *Aviation Week & Space Technology* (17 October 1983): 16.
2. "Draft Treaty on the Prohibition of the Stationing of Weapons of Any Kind in Outer Space," submitted 11 August 1981.
3. Article 2 of the Soviet draft did call upon all nations to use satellites "in strict accordance with international law," and in the past the Soviets at various times had asserted that photo surveillance and direct-broadcast satellites, as well as the U.S. space shuttle, violated international law. Yet the wording and structure of the Soviet draft indicates that Article 2 was more hortatory than a formal obligation,

and the obligation in Article 1 not to attack satellites would remain binding, even if those satellites violated international law—provided they did not carry "weapons."

4. *Aviation Week & Space Technology* (8 February 1982): 21.
5. White House Fact Sheet, "National Space Policy," 4 July 1982, declassified summary of Reagan's Presidential Directive on National Space Policy, p. 4.

   On the formulation of the ASAT target list, see the testimony of Gen. Thomas Stafford, Air Force Deputy Chief of Staff for Research, 27 March 1979; U.S. Senate, Committee on Armed Services, *Hearings, Department of Defense Authorization for Appropriation for Fiscal Year 1980,* 96th Cong., 1st sess., Part 6, p. 3037.

   U.S. plans to use its ASAT to intercept the Soviet ASAT have been reported in the press. See, for example, Thomas Karas, *The New High Ground* (New York: Simon & Schuster, 1983), p. 157, and *Aviation Week & Space Technology* (8 February 1982): 21. But clear public statements by official spokesmen on this point are hard to find, and a recent Government Accounting Office study reportedly criticized the U.S. ASAT for its inability to attack the Soviet ASAT. See *Aviation Week & Space Technology* (19 December 1983): 21.
6. *Aviation Week & Space Technology* (19 December 1983): 21–22.
7. See Report by the Comptroller General of the United States, "U.S. Antisatellite Program Needs a Fresh Look," unclassified digest of a classified report, 27 January 1983.
8. The Air Force's Space Defense Operations Center (SPADOC) relies upon widely scattered radar and special telescope sites to track Soviet and U.S. spacecraft. The radars include three ballistic missile early warning system radars in Alaska, Greenland, and Great Britain facing north toward the Pole; one radar in North Dakota (formerly part of the Safeguard ABM system) facing north; four warning radars in the continental United States facing out over the Atlantic, Pacific, and Gulf of Mexico to detect SLBM attacks; two monitoring radars in Turkey and the Aleutians facing toward Soviet missile test ranges; Navy space surveillance system radars in six states along the southern rim of the United States; and three other spacetrack radars on Antigua and Ascension islands in the Caribbean and South Atlantic and on Hawaii. Radars will be added or upgraded in the Philippines, Guam, and Kwajalein facing north over the western Pacific. Upgraded radars in Turkey, Kwajelein, and Massachusetts will be able to track objects at geosynchronous orbits. The special telescopes are located in Canada, California, Hawaii, Italy, Korea, New Mexico, and New Zealand. These can only track objects against the night sky, and the amount of the sky each can scan in one night is limited. They will be augmented

or replaced in some cases by the GEODSS network. See Thomas Karas, pp. 35–36; *Aviation Week & Space Technology* (22 February 1982): 65; and note 13 below.

9. For arguments along these lines, see the testimony of Eugene Rostow, then-director of the Arms Control and Disarmament Agency, U.S. Senate, Committee on Foreign Relations, *Hearings, Arms Control and the Militarization of Space,* 97th Cong., 2nd sess., 20 September 1982; and State Department news briefing, 26 August 1983, in John Pike, "Limits on Space Weapons" (Washington: Federation of American Scientists, 1983): 17.

10. See Nicholas L. Johnson, *The Soviet Year in Space, 1982.* (Colorado Springs, Colo.: Teledyne-Brown Engineering, 1982): 26; Robert Berman & John Baker, *Soviet Strategic Forces* (Washington, D.C.: The Brookings Institution, 1982), p. 152; and *Aviation Week & Space Technology* (28 June 1982): 20–21, for useful summaries of Soviet ASAT tests through June 1982.

    The Soviet resumption of ASAT tests in February 1976 after a four-year hiatus, has been attributed by some as "a warning to the Chinese, who launched their first reconnaissance satellite in September 1975." [Marcia S. Smith, "Antisatellites (Killer Satellites)" Issue Brief Number IB 81123, Science Policy Research Division, Congressional Research Service, Library of Congress, August 1981, p. 4.] The pattern of Soviet ASAT launches within the last decade does parallel China's satellite program. It had been widely anticipated in the West that China would launch its first satellite in 1968, the same year that Soviet ASAT tests began. Instead, the first two Chinese satellite launches occurred on 24 April 1970 and 3 March 1971, followed then by a pause of four years. The last Soviet ASAT test before their own hiatus was in December 1971. The Soviets put another target satellite up in September 1972, but this target was evidently never attacked and conceivably it was a test of the target system only. It seems more probable, however, that the Soviets did not decide to discontinue ASAT tests until around this time—that is, at a point when more than a year of Chinese nonactivity in space had passed. The Chinese then resumed their satellite program with launches on 26 July, 26 November, and 16 December 1975, involving large spacecraft weighing several tons. The 26 November launch was notable, because for the first time the Chinese reportedly separated and recovered a capsule from a satellite—a strong indicator that this satellite had a photoreconnaissance mission. The capsule recovery occurred on 2 December 1975. The Soviets then broke their hiatus, launched a new target, and carried out an ASAT intercept in the second week of February 1976.

11. "Draft Treaty on Banning the Use of Force in Space and from Space with respect to the Earth", Article 2(4); submitted 18 August 1983.

12. It is often argued that nuclear-armed ICBMs could perform reliably as ASATs during conflict, even if never tested in simulated ASAT attacks. The large kill radii of nuclear warheads is assumed to compensate for other shortcomings in guidance and missile performance. It is hard to argue otherwise, because few if any operational ICBMs seem ever to have been launched on trajectories simulating high-orbit ASAT engagements. The reliability of ICBMs when fired on trajectories other than the ones to which they have been tested, however, has been a subject of much debate recently in the context of controversies over land-based ICBM vulnerability.
13. *Aviation Week & Space Technology* (28 February 1983): 57–58.
14. *Aviation Week & Space Technology* (22 February 1982).
15. Because of the general difficulties of doing such calculations and adjustments, NASA in the early days of manned space flight simply did not bother and trusted instead to the emptiness of space as the best insurance against collisions. The Air Force's Space Defense Operations Center (SPADOC) will acquire a new computer by 1985 to perform such calculations. *Aviation Week & Space Technology* (8 February 1982).

REFERENCES

Berman, Robert, and John Baker. *Soviet Strategic Forces.* Washington, D.C.: The Brookings Institution, 1982.

Covault, Craig. "Space Defense Organization Advances." *Aviation Week & Space Technology (AWST)* (8 February 1982): 21.

———. "USAF Awaits Space Defense Guidance." *AWST* (22 February 1982): 65.

"Draft Treaty on Banning the Use of Force in and from Space with Respect to the Earth." *Pravda* (23 August 1983). In U.S. Foreign Broadcast Information Service *Daily Report—Soviet Union* 163 (22 August 1983): AA2–AA4.

Johnson, Nicholas L. *The Soviet Year in Space: 1982.* Colorado Springs, Colo.: Teledyne-Brown Engineering, 1982.

Karas, Thomas. *The New High Ground.* New York: Simon & Schuster, 1983.

Pike, John. "Limits on Space Weapons." Washington, D.C.: Federation of American Scientists, 1983.

Randolph, Anne. "USAF Upgrades Deep Space Coverage." *AWST* (28 February 1983): 57–58.

Robinson, Clarence. "Beam Weapon Advances Emerge." *AWST* (18 July 1983): 18–21.

———. "Panel Urges Defense Technology Advances." *AWST* (17 October 1983): 16.

———. "USAF Will Begin ASAT Testing." *AWST* (19 December 1983): 21.

Smith, Marcia S. "Antisatellites (Killer Satellites)." Issue Brief Number IB 81123, Science Policy Research Division, Congressional Research Service, Library of Congress, August 1981.

"Soviets Stage Integrated Test of Weapons." *AWST* (28 June 1982): 20–21.

United Nations. General Assembly. "Draft Treaty on the Prohibition of the Stationing of Weapons of Any Kind in Outer Space." Submitted by the USSR 11 August 1981. A/RES/36/97 (15 January 1982).

U.S. Congress. General Accounting Office. "U.S. Antisatellite Program Needs a Fresh Look." Unclassified digest of a classified report, 27 January 1983.

———. Senate. Committee on Armed Services. *Hearings on Department of Defense Authorization for Appropriations for FY 1980, Part 6.* 96th Cong., 1st sess. 29 March 1979. Testimony of Gen. Thomas P. Stafford: 3037.

———. Senate. Committee on Foreign Relations. *Hearings on Arms Control and the Militarization of Space.* 97th Cong., 2d sess. 20 September 1982. Testimony of Hon. Eugene Rostow: 11–12.

U.S. Department of State. *Department of State Bulletin.* 26 August 1983.

The White House. "Fact Sheet—National Space Policy." *Weekly Compilation of Presidential Documents* 18 (18 July 1982): 872–76.

INDEX

ABOUT THE EDITOR

William J. Durch is currently a research fellow at the Center for Science and International Affairs at Harvard University's John F. Kennedy School of Government. Before joining the staff of CSIA in 1981, he was a Foreign Affairs Officer with the U.S. Arms Control and Disarmament Agency in Washington, D.C., where he served on the Policy Planning Staff and worked in the areas of strategic defense and space defense. From 1973 to 1978 he was on the staff of the Center for Naval Analyses in Alexandria, Virginia. He holds a BSFS from the Georgetown University School of Foreign Service and an MA in political science from George Washington University.

ABOUT THE CONTRIBUTORS

Donald L. Hafner is Associate Professor of Political Science at Boston College. He served as an advisor with the U.S. SALT delegation and worked on antisatellite issues for the National Security Council ASAT Working Group, 1977–78. He holds a doctorate in political science from the University of Chicago.

Louise Hodgden is a Graduate Research Associate at the Center for Science and International Affairs (CSIA), Harvard University, and a PhD candidate at the Fletcher School of Law and Diplomacy.

George F. Jelen is a Research Fellow at the Harvard Program on Information Resources Policy (HPIRP) and an Adjunct Research Fellow, CSIA. He is the author of the forthcoming HPIRP study, *Information Security: Elusive Goal,* and holds an MS in Technology of Management from the American University.

Stephen M. Meyer is Associate Professor of Political Science at the Massachusetts Institute of Technology and director of the Soviet Security Studies Working Group at MIT's Center for International Studies. He has a doctorate in political science from the University of Michigan.

Philip D. O'Neill, Jr. is a partner with the law firm of Hale and Dorr and an Adjunct Research Fellow at CSIA. He holds a JD from Boston College School of Law.

Malcolm Russell is an associate with the law firm of Mintz, Levin, Cohn, Ferris, Glovsky and Pope. He holds a JD from Harvard Law School and an MA in Russian Studies from Yale University.

Paul B. Stares is a Rockefeller International Relations Fellow in residence as a Guest Scholar at the Brookings Institution. He has recently completed a doctoral dissertation on the evolution of the U.S. antisatellite program. His doctorate in political science is from the University of Lancaster, England.

Dean A. Wilkening is a Postdoctoral Research Fellow at CSIA. He holds a doctorate in physics from Harvard University.